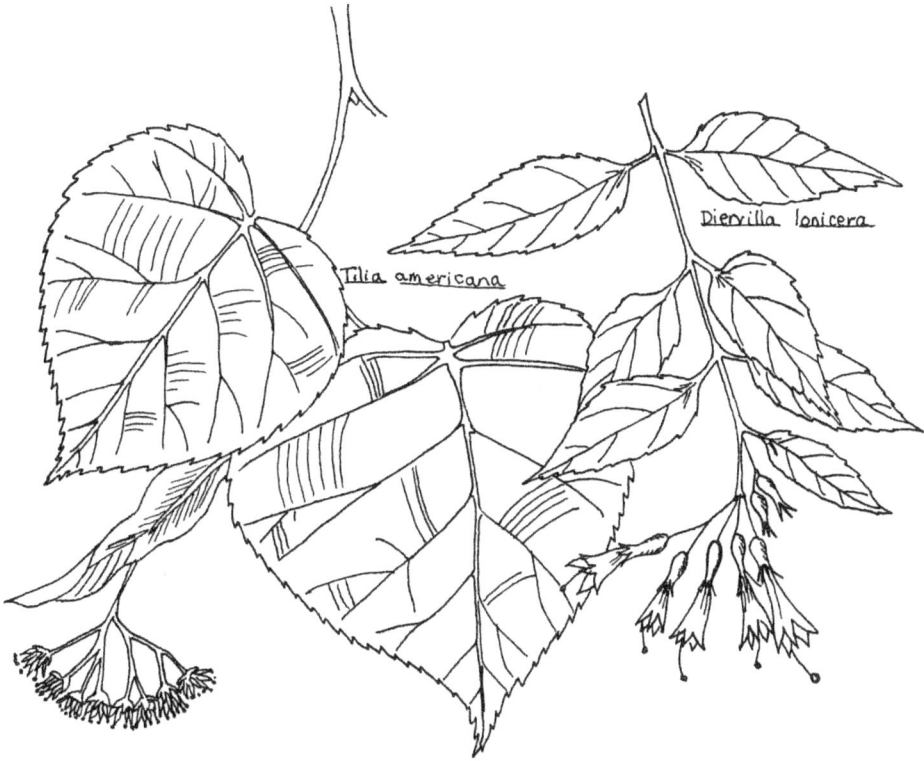

Tilia americana
Diervilla lonicera

Plants for Beekeeping In Canada and the Northern USA

A Directory of Nectar and Pollen Sources Found in Canada and the Northern USA

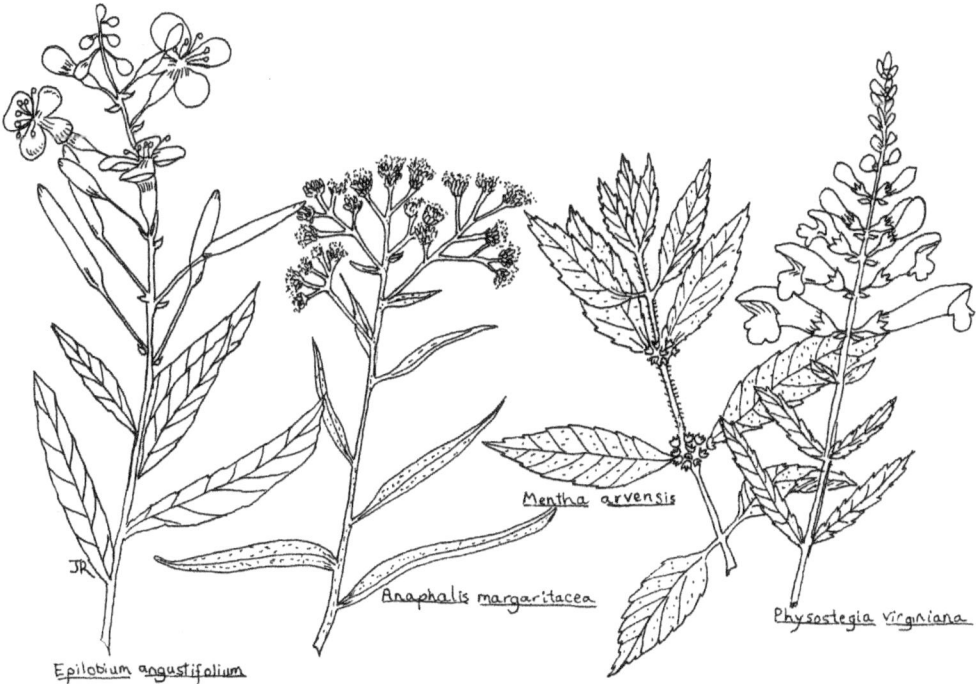

Mentha arvensis
Anaphalis margaritacea
Physostegia virginiana
Epilobium angustifolium

REVISED EDITIONS
First printed by Burlington Press, Foxton, Cambridge, UK (1987).
Front cover photo by Damon Hart-Davis.
This edition printed and published by Northern Bee Books 2025.

ISBN: 978-1-914934-96-4

Plants for Beekeeping In Canada and the Northern USA

A Directory of Nectar and Pollen Sources
Found in Canada and the Northern USA

by Jane Ramsay

TABLE OF CONTENTS   Page

List of Tables

List of Figures

Robinia pseudoacacia

FOREWORD

The compilation of this important reference book on nectar and pollen sources has been an ambitious project Canada is the fifth or sixth leading honey-producing country in the world, and has a very wide array of forage sources for honeybees. Yet this is the first time that the extensive literature available on the native and introduced plants of Canada and the northern USA has been surveyed from the point of view of beekeeping – and the information published under one cover.

Jane Ramsay has identified over 450 plants which have a high honey potential, or provide useful sources of pollen, and gives details about where these plants will grow in Canada, when they bloom, and the nature of the honey that they produce. She presents the information for each plant under standard headings – based on those in the IBRA publication 'A Directory of Important World Honey Sources' – so that readers can quickly locate details of a particular plant and compare entries for different plants.

I hope that *Plants for Beekeeping in Canada* may indirectly contribute towards increasing honey production in the region. Beekeepers who are interested in fixed-land honey production – honey from plants which are selected for cultivation purely for their use as bee forage – will find much useful information herein. The book shows also where there are gaps in the published literature, and I hope that it will prompt beekeepers and bee researchers to provide the observations and data that will fill in these gaps.

The author is particularly well qualified to compile this book, having trained in Canada both as an apiculturalist and an 'agrologist'. She has written a book that will appeal to a wide readership, ranging from those who know a great deal about beekeeping and bee forage plants to those who know very little but are keen to learn more on this fascinating subject.

Margaret Adey,
IBRA, Cardiff, UK,
March, 1987

* * *

The preparation of this book was an ambitious project and was initiated while the author was attending Simon Fraser University, and this is where we first met. Jane identifies over 450 plants which grow in Canada and have nectar producing potential. There is also considerable information on a variety of related topics, including ways to improve the attractiveness of plantings to bees. *Plants for Beekeeping* is an important reference book for beekeepers and others who want useful information on nectar and pollen plants. Canada is a major honey producing country and yet, prior to this work, there was no extensive literature available.

Douglas M. McCutcheon,
author of *A History of Beekeeping in British Columbia*,
Langley, BC,
February 2015

PREFACE

The first purpose of this book is to compile useful information on bee forage and present it in such a way that published facts about a nectar or pollen source can be found easily and quickly. The second and more important purpose is to find out which plants deserve more research or observation regarding their nectar or pollen potential. For instance, certain plants such as conifers (for honeydew), goldenrods, fireweeds and asters have not been studied thoroughly, but are known to be significant contributors to surplus honey production in Canada. Many other less well-known plants probably warrant more appreciation and a closer look.

This book is written at a time when the Canadian beekeeping industry is confronted with two crises: acarine and varroa pests and the Africanized honeybee that will spread into at least the southern USA, disrupting beekeeping and the supply of package bees and queens to Canada.

These are serious problems and apiary management practices are being adapted so that the Canadian beekeeping industry will continue to prosper. I am optimistic that the beekeeping industry will survive the present crises and continue to be one of the most successful in the world. Bee plants are an essential ingredient of that success; in them lie the hope of bountiful honey harvests for years and years to come.

Jane Carver Ramsay,
Cambridge UK, December, 1986
(edited February, 2015)

* * *

After 30 years it seems appropriate to edit the first preface and add a postscript. First and foremost, I am so grateful for the continued interest of readers. It is also comforting to know that other people still read books. Much has happened since 1986 pertinent to apiculture and agriculture in general. Of major concern is a decline in pollination of some crops and native plants. This has drawn attention to the value of honeybees and native pollinators, including native bees. My hope is that this book will also prove useful to those who wish to grow plants of benefit to all pollinators. Preserving or creating large and small areas of wildness, rich with native flora, will offer habitat and food for native bees and other beneficial insects. Nectar and pollen corridors will help support migratory pollinators and be helpful to many species. I plan to spend some of my retirement free time getting to know my native bee neighbours and planting for them.

Jane Carver MacDonald,
Victoria BC,
February, 2015

ACKNOWLEDGEMENTS

First of all, I must thank the International Bee Research Association (IBRA) for being constantly supportive and offering me a great deal of editorial and administrative help during the production of this book. Without IBRA's encouragement I know that it would have been difficult or impossible to finish. IBRA was able to offer only limited support for publication because of financial constraints. Nevertheless, the Association believed that this book would be valuable to beekeepers and apiculturalists in Canada and the northern USA. Therefore the Association determined to formally support publication and offered whatever help it could to assist with production.

I am grateful for the generous help and advice of many friends and colleagues. Without their assistance this book would never have been written, edited and finally printed. I can only mention some of those kind and talented people who have helped.

My sincere thanks go to Douglas McCutcheon, Supervisor of Apiculture Programs in British Columbia, who has long wished to have a reference of this kind for his province. He should be credited as the source of the inspiring idea behind this book. Not only that, he has offered practical help by loaning me a good deal of his valuable reference material, circulating a draft copy to several people for comments, and finding support for the book among local bee associations which then made financial contributions towards advertising and distribution costs.

Dr. Mark Winston of Simon Fraser University (S.F.U.) has always been helpful, and is one of those people who has always believed that I would eventually finish. My thanks go to Steve Mitchell and Kerry Clark, both formerly of S.F.U., for carefully reading through and commenting on the manuscript. Also to Doug Colter, Chief Apiary Inspector for Alberta, for allowing me to use Table 14. I am grateful to the Prince Edward Island (P.E.I.) Beekeeper's Cooperative Association for providing a typist to prepare an early draft of the manuscript. I would like to thank Libby Oulton of Ragweed Press in Charlottetown for her enthusiasm and her attempts to find a Canadian publisher. Keith Brehaut of Bunbury Nursery in P.E.I. gave practical suggestions that helped to improve Section 7. I am grateful to Harry Baglole who was honest enough to tell me that my writing style in the first draft could do with some improvement. Our neighbour and friend, Patrick Miles, has given me useful professional advice on publishing and printing.

I spent several very pleasant and productive weeks using the excellent Library of the University of Cambridge's Botanic Garden. I am infinitely grateful to Mr. Clive King, Librarian and Taxonomist at the Botanic Garden, for his patience and indispensible assistance in finding references that I needed for checking Latin names and botanical details.

Dr. Sarah Corbett, of the Applied Biology Department at the University of Cambridge, I cannot thank enough for her keen interest in the project (in spite of the fact that it promotes forage for honeybees rather than "wild" bees which are her chief interest). She also introduced me to the staff of the Computer Services Department at the University of Cambridge, who have made processing of the text possible by offering me the use of their excellent facilities. My very sincere gratitude is given to the Computer Services Department. Alison Emery, a hard working typist (in her spare time), spent many hours entering text into the mainframe. Mr. Roger Smith of Ozymandius Productions (who understands the mysteries of computers) skillfully worked on layout, the index, the tables and figures, and many other last minute problems. His expert and timely help has been invaluable.

There are a number of staff at IBRA who deserve a very warm thank you for their assistance in checking references, writing for permissions, offering useful suggestions, and of course for including the book in their list. Special thanks go to Karl Shower, Administrative Officer of IBRA, and Penelope Walker (formerly of IBRA). Dr. Margaret Adey, Director of IBRA, has been constantly supportive and helpful, often giving me her time on her precious week-ends. The past Director, Dr. Eva Crane, should be given full credit for the idea of including the section on Canadian beekeeping, and for her ideas about slide illustrations which may be incorporated into a second edition.

Jane Ramsay,
Cambridge, UK
April, 1997

* * *

The small changes included in this revised edition were made with the help and care of book designer, Iryna Spica and with the support of Printorium Bookworks. I am extremely grateful for Iryna's patience and expertise.

My long time friend, Doug McCutcheon, kindly agreed to write a new foreward and I am very honoured. Many years ago, when I was a student and Doug was the Supervisor of Apiculture Programs in BC, he was very generous with his time and kindly shared his plant knowledge and files with me. It was his help that started me on this book and I dedicate this modestly revised edition to Doug McCutcheon in gratitude and in admiration of his recent book, A History of Beekeeping in British Columbia.

Jane Carver MacDonald,
Victoria, B.C.
February, 2015

1. INTRODUCTION

"Late in March, when the days are growing longer
     And the sight of early green
Tells of the coming spring and suns grow stronger,
     Round pale Willow-catkins there are seen
          The first year's honey-bees
Stealing nectar; and the bee-masters know
This for the first sign of the honey-flow. ..."

Martin Armstrong, *Honey Harvest*

This book is a reference for beekeepers, apiculturalists, agriculturalists, botanists and gardeners who wish to know which plants are attractive to honeybees, where they will grow in Canada, when they will bloom and the nature of the honey produced. It is written so as to be comprehensible to a wide readership. Unnecessary jargon has been excluded and technical terms are explained.

The main body of the book consists of descriptions of over 450 plants arranged horticulturally into three sections:

annuals and biennials
perennials
trees and shrubs

The information describing the plants appears in the following order in each case:

scientific name
common name or names
horticultural designation
hardiness rating
blooming period
height range
whether nectar, pollen or honeydew collected
Value for honey, including: honey potential in kg/ha; nectar sugar concentration; honey yield per colony; nectar flow characteristics
Value for pollen, including: quality, quantity; attractiveness
Honey, including: colour; granulation density; flavour; aroma
Notes, including: range within Canada and the adjacent northern USA.

The supporting text provides background information on the subject of nectar and pollen sources, including: historical notes; the need for bee forage; nectar and pollen; planting for continuous bloom; unpalatable honey; poisonous plants; honeydew; pesticides; and Canadian beekeeping. The tables and lists of plants are designed to assist the reader in finding information quickly and easily. These include: early pollen sources; highly attractive nectar plants; plants with potential for roadsides and banks; and calendars of plants.

2. BACKGROUND

For centuries beekeepers have dreamed of a "wonder honey plant", a plant worth growing solely for the value of the honey crop it yields. A number of plants have been given names which suggest that they might be this beekeepers' cornucopia. Most of these species acquired their evocative common names during Victorian times, and these names are still the ones most frequently used:

golden honey plant	(*Verbesina alternifolia*)
wonder honey plant	(*Agastache foeniculum*)
Rocky Mountain bee plant	(*Cleome serrulata*)
Chapman honey plant	(*Echinops sphaerocephalus*)
the bee tree	(*Evodea daniellii*)
Simpson's honey plant	(*Scrophularia marilandica*)
honey locust	(*Gleditsia triacanthos*)

Although some of these are indeed good bee plants, most have been greatly overrated. However, exaggerated or not, the enthusiasm of the Victorian beekeeper-entrepreneurs is infectious and serves as an inspiring tonic. The following quotation is typically optimistic.

"It [*Agastache foeniculum*[a], wonder honey plant[b]] produces honey in the greatest abundance, which possesses in a slight degree the same fragrance as the plant, which renders it exceedingly pleasant to the taste... I firmly believe that an acre of this plant well established would be ample pasturage for 100 colonies of bees.."

Beekeeper's Journal, 1872[72].

Interest in growing bee plants was indeed high during the late nineteenth century in North America, when beekeeping was a more sedate and less mechanized occupation. Migratory beekeeping was not usually physically possible, so the prospect of providing improved bee forage near apiaries led to experimentation on a large scale. Candidate bee plants were sown in "test gardens", fields and roadside verges.

Today, the pattern of beekeeping life is different, less labour intensive but perhaps at the same time more hectic. The short honey season is governed by consecutive deadlines. Entire beeyards are now commonly moved from one location to the next, following the nectar flows. One reason beekeepers need to move yards is that there has been a marked decline in the availability and variety of bee forage. This has been caused by urban expansion and agricultural intensification. Both activities have reduced the diversity of native bee plant species by "tidying up" the landscape with bulldozers and herbicides.

Fortunately, there seems to be a renewed interest in growing plants which are attractive food sources for bees. Gardeners and beekeepers can derive a great deal of pleasure from planting a garden that offers a continuous and assured food supply from

---

[a] The scientific name is not specified in the original reference. The description may instead apply to *Agastache nepetoides*, yellow giant hyssop, found in Que. and s. Ont. and S. into the USA.

[b] Other common names for this species are: anise hyssop, blue giant hyssop, fennel giant hyssop, fragrant giant hyssop.

crocus to aster. Not only is such an enterprise practical, providing bee forage when there is little available, but it is also creative and interesting. Gardeners especially will enjoy becoming acquainted (or re-acquainted) with many of the "old fashioned" garden plants that are often the best bee plants.

However, the honey industry is now mainly dependent on commercial agricultural crops for economic yields. Since ultimately nectar and pollen production are genetically determined, and may be reduced or increased as crops are engineered to perform for the market place, beekeepers are often in the hands of plant breeders. In some cases nectar production cannot be ignored by plant geneticists, for instance in crops that are grown for seed and require pollination by honeybees. Beekeepers should continue to try to influence plant breeding research so that it benefits the honey industry, and in this way help to protect their own livelihood.

Planting for wild bees should not be overlooked. Our native bees such as bumblebees, *Bombus*, make a significant contribution to commercial agriculture since they are often more efficient pollinators than honeybees (e.g. for many commercially grown cherries). Like honeybees, wild bees are also suffering from diminished resources. Wild (native) species are severely threatened by habitat destruction and the consequent lack of suitable nest sites. Undisturbed "conservation" areas at the bottom of gardens or in vacant urban lots can purposely be left to become enticing locations for nesting bumblebees and other wild bees. Also, some plants attractive to honeybees are visited by their wild relations. Increasing honeybee forage will therefore sometimes benefit wild bee species by providing them with more food sources.

In these small but important ways local populations of wild bees could be encouraged and protected. In any case, more bees of all kinds in the garden and orchard will result in larger harvests of better quality produce, so there is a promise of mutual benefit when we share our gardens with bees.

Some of the more important nectar and pollen producing plants in Canada are shown in Table 1. The list includes both naturally occurring and cultivated plants. Beekeepers could note which of these are known locally as important sources, when they bloom each year and when the major gaps occur in the nectar flow. They could then consider planting bee forage to help fill the dearth periods. Table 18 is more general, but gives the approximate flowering times of ten major nectar and pollen sources in different beekeeping regions of Canada.

TABLE 1. Some naturally occurring plants and cultivated crops which are important nectar and pollen sources in Canada.
(Adapted from Robinson and Oertel[78]. Revised by D. McCutcheon, 1986).
N = nectar
P = pollen

Plant	Eastern Canada*	Western Canada*
alder (*Alnus* spp.)	P	
alfalfa (*Medicago sativa*)		NP
aster (*Aster* spp.)	NP	NP
basswood (*Tilia americana*)	NP	N
bird's foot trefoil (*Lotus corniculatus*)	P	P
blackberry (*Rubus* spp.)	NP	NP
blueberry (*Vaccinium* spp.)	N	NP
blueweed (*Echium vulgare*)	NP	
boneset (*Eupatorium* spp.)	N	
buckthorn (*Rhamnus* spp.)	NP	NP
buckwheat (*Fagopyrum esculentum*)	N	N
cascara (*Rhamnus purshiana*)	N	N
chicory (*Cichorium intybus*)	NP	NP
clover		
alsike (*Trifolium hybridium*)	NP	NP
Persian (*Trifolium resupinatum*)	NP	NP
red (*Trifolium pratense*)	NP	NP
sweet (*Melilotus* spp.)	NP	NP
white (*Trifolium repens*)	NP	NP
coneflower (*Rudbeckia* spp.)	NP	NP
corn (*Zea mays*)	P	
dandelion (*Taraxacum* spp.)	NP	NP
dogbane (*Apocynum androsaemifolium*)	N	N
elderberry (*Sambucus canadensis*)	P	
elm (*Ulmus* spp.)	NP	
fireweed (*Epilobium angustifolium*)	NP	NP
fruit bloom, trees		
apple (*Malus* spp.)	NP	NP
cherry (*Prunus* spp.)	NP	NP
plum (*Prunus* spp.)	NP	NP
gaylussacia (*Gaylussacia* spp.)	N	
giant hyssop (*Agastache nepetoides*)	NP	NP
goldenrod (*Solidago* spp.)	NP	NP
hawthorn (*Crataegus* spp.)	NP	NP
hound's tongue (*Cynoglossum officinale*)	N	
knapweed (*Centaurea repens*)	NP	NP
Labrador tea (*Ledum groenlandicum*)	N	N

TABLE 1. Continued

Plant	Eastern Canada*	Western Canada*
locust, black (*Robinia pseudoacacia*)	NP	NP
maple (*Acer* spp.)	NP	NP
milkweed (*Asclepias* spp.)	NP	NP
mustard (*Brassica* spp.)	NP	NP
oak (*Quercus* spp.)	P	
poplar (*Populus* spp.)	P	P
privet (*Ligustrum* spp.)	NP	
ragweed (*Ambrosia* spp.)	P	P
rape, canola (*Brassica* spp.)	NP	NP
raspberry (*Rubus* spp.)	NP	NP
Russian thistle (*Salsola* spp.)		NP
safflower (*Carthamnus tinctorius*)	NP	NP
sainfoin (*Onobrychis* spp.)		NP
sheep laurel (*Kalmia angustifolia*)	NP	
Siberian almond (*Pyrus baccata*)**	NP	
smartweed (*Polygonum* spp.)	NP	N
snowberry (*Symphoricarpos occidentalis*)		NP
sumac (*Rhus typhina*)	NP	NP
sunflower (*Helianthus* spp.)		NP
tarweed (*Hemizonia* spp.)	NP	
thistles (*Cirsium* spp.)	NP	NP
vetch (*Vicia* spp.)	NP	NP
viper's bugloss (*Echium vulgare*)		NP
wild alfalfa (*Lotus* spp.)	NP	
wild radish (*Raphanus raphanistrum*)	NP	NP
willow (*Salix* spp.)	NP	NP

* In the original table by Robinson and Oertel[78], eastern Canada is represented only by New Brunswick, Ontario, and Quebec. However many of these plants are also important nectar sources in Nova Scotia and Prince Edward Island. The western provinces are represented by Alberta, Manitoba, Saskatchewan and British Columbia.
** This is usually known as Siberian crab apple (*Malus baccata*).

3. NECTAR AND POLLEN:
WHY PLANTS ARE ATTRACTIVE TO BEES

Plants vary in their attractiveness to bees. About 90% of Canadian honey comes from less than 3,000 or so plant species visited by bees in North America[78]. The notes below may help to explain why the dominance of such a small group of plants is not due to their acreage alone.

3.1 Nectar

Nectar is the raw material used by bees to manufacture honey, the individual's or colony's source of energy. It is essentially a solution of sucrose and water which is secreted by plants from special glands called nectaries. The nectaries are usually located at the base of the flower petals. In some species nectaries are found on other parts of the plants such as leaves or stems (e.g. *Vicia*, vetch). These are termed extra-floral nectaries.

Honeybees collect nectar by inserting their tongues into the blossom near the nectary (or into the extra-floral nectary) and sucking the nectar up into a nectar or honey sac inside their bodies. When a bee has a full load she returns to the colony and passes her nectar to other bees which deposit it in cells. While the nectar is in a bee's honey sac and during the transfer to other bees most of the sugars are converted into dextrose and fructose (inversion) and some of the water in the nectar is evaporated (ripening). The ripened and inverted nectar, now honey, is stored in cells which are then capped over with wax for use later. The unique flavour and aroma of honey from a single floral source is largely due to the aromatic oils in the nectar.

Nectars with high sugar concentrations are particularly attractive to bees because their reward for the energy expended in foraging is greater. The average concentration of sugar in the nectar of species visited by honeybees ranges from 20-40%[92]. When there is little forage available bees will generally collect nectars of lower concentrations than they would during a honey flow. Table 2 lists some very good bee plants in order of decreasing nectar concentration. The sequence is of more significance than is the actual concentration figure.

The total volume of nectar secreted is just as important as the amount of sugar in the nectar. For instance, *Prunus* (e.g. cherry, plum) produce a very dilute nectar (approximately 15% on average), but because the quantity produced in a 24-hour period is very high they are able to attract bees. The weight of sugar produced over 24 hours has been measured for a number of plants, and is a factor used to estimate the honey potential (HP) value given for the plants described in this book (see Section 4.7 for an explanation of HP).

Another significant factor that is important in determining the value of a plant as a honey source is the depth of its blossoms (corolla length). Blossoms which are about 6 mm or less in depth are well suited to honeybees because all, or nearly all, the nectar is within reach of a honeybee's tongue. Nectar offered in blossoms which are greater than 6 mm deep is mostly out of reach of the honeybee. For instance, most honeysuckles (*Lonicera*) have very deep blossoms which are suited to the much longer tongue of the bumblebee, and are seldom visited by honeybees. There are some families of plants which are particularly rich in attractive bee flora, two outstanding ones being the *Labiatae* (mints, thymes, sages, etc.) and the *Leguminosae* (pulses clovers, alfalfa etc.). It is not surprising that many of the labiates and legumes have blossoms especially well adapted in shape and depth to honeybee anatomy and behaviour.

Double blossomed cultivars, although visually very attractive, usually produce little or no nectar and pollen. This is because the petal number and blossom size have been increased for show at the expense of the sexual parts of the flower.

TABLE 2. Average sugar concentration in the nectar of some plants commonly visited by honeybees, arranged in decreasing order (measured in Colorado; from Wilson *et al.*[94]).

Scientific name	% sugar	Common name
Rosa spp.	58.4	rose
Symphoricarpos occidentalis	57.2	snowberry
Trifolium pratense	54.8	red clover
Brassica spp.	51.5	rape, mustard etc.
Trifolium hybridum	51.4	alsike clover
Malus pumila	50.3	apple
Helianthus petiolaris	49.2	Prairie sunflower
Trifolium repens	48.8	white Dutch clover
Melilotus officinalis	48.7	yellow sweet clover
Medicago sativa	44.9	alfalfa
Lonicera tatarica	44.7	Tatarian honeysuckle
Melilotus alba	44.3	white sweet clover
Cucumis sativa	42.2	cucumber
Aster laevis	40.9	smooth aster
Taraxacum officinale	36.1	dandelion
Alcea rosea	33.7	hollyhock
Solidago occidentalis	32.8	western goldenrod
Cichorium intybus	32.6	chicory
Gaillardia spp.	32.0	gaillardia
Solidago canadensis	31.4	Canadian goldenrod
Cleome serrulata	29.7	Rocky Mountain bee plant
Melilotus alba 'Annua'	29.6	Hubam clover
Trifolium incarnatum	29.2	crimson clover
Aster novae-angliae	24.6	New England aster
Nepeta cataria	22.3	catnip, catmint

3.2 Pollen

A colony's future depends as much on its pollen stores as on its honey reserves, since pollen is an essential part of the diet of young bees. Without pollen a colony cannot grow and develop. Chemically, pollen largely consists of amino acids (protein), vitamins, and minerals. It is produced in the anthers or male sexual parts of flowers and must be transferred by bees, other insects, birds, bats or wind (depending on the plant) to the female part of the flower for pollination to occur.

While foraging for nectar and pollen a bee usually becomes dusted or daubed with pollen which adheres to her plumed body hair. If a bee is actively collecting pollen, she moistens the grains and packs them into the two pollen baskets on her hind legs for the flight back to the hive. The pollen is then stored in open cells, pollen from different sources often being stored in different cells. Since pollens range in colour from deep red through every colour of the rainbow to blue, a well-filled comb can look rather like a stained glass window. Beekeepers will often know what plants bees are working by the colour of the pollen load on a bee's legs. Precise identification of the source of a particular pollen requires examination of the individual pollen grains using a microscope.

Most beekeepers recognize that it is vital to have an ample winter store of high quality pollen in the hive in late winter or early spring so that brood rearing will start before pollen supplies are available in the field. However, even with sufficient stores, it is still extremely important to have native or cultivated pollen-producing plants blooming near colonies in the early spring. Freshly collected pollen is more nutritious than stored pollen, and seems to be more effective in stimulating a queen to begin laying eggs. Some dependable early pollen sources include *Crocus* (crocus), *Prunus* (cherry), and *Salix* (willow). A more extensive list can be found in Section 13.1. Many plants provide pollen for honeybees, but the quantity, quality and ease of collection vary greatly from one plant to another. Pollens range in protein content from about 10 - 30%[83], and so also vary in their nutritive value to honeybees (Table 3). Certain plants such as poppy, pear and hollyhock produce a pollen which is relished by bees. It seems that the attractiveness of some pollens may be due to factors other than protein content.

TABLE 3. A few examples of plants with different pollen protein levels (adapted from Stanley et al.[83]).

Pollen protein level	Generic name	Common name
excellent	Castanea	chestnut
	Crocus	crocus
	Erica	heath
	Papaver	poppy
	Prunus	cherry
	Salix	willow
	Sinapis	mustard
	Trifolium	true clover
good	Taraxacum	dandelion
	Ulmus	elm
poor	Alnus	alder
	Betulus	birch
	Corylus	hazel
	Populus	poplar
very poor	Abies	fir
	Cedrus	cedar
	Picea	spruce
	Pinus	pine

4. THE PLANTS

4.1 Scientific and common names

Botanists use scientific names to ensure positive identification of a plant, and they are used here for the same reason. A scientific or botanical name consists of two words, the generic name followed by the specific name, and is always underlined or italicized (e.g. *Acer rubrum*). The generic name may be abbreviated by giving the first letter only (e.g. *A. rubrum*).

When a plant is re-evaluated and perhaps re-classified by botanists, its scientific name will probably be changed in order to define its identity more accurately. As a result, both the former and the latest accepted scientific names are often used interchangeably, at least for a few years until the contemporary name becomes established.

Here, a diligent attempt has been made to keep the nomenclature consistent. This is achieved by using names given in the most recent edition of a well known North American botanical encyclopaedia, Hortus Third[49]. This reference is used exclusively, except in a few cases where another was needed due to apparent disagreement in the literature about names, and also because of evidence of name changes subsequent to the publication of Hortus Third. In these few cases, the botanical work that is referred to is cited in the text.

Whenever it is thought to be helpful, the former synonymous scientific names are given in brackets below the contemporary, "correct" name. For example:

Alcea rosea L.
(syn. *Althaea rosea* L.).

All scientific names, former and contemporary, are listed in the Index of Scientific Names.

The method for naming follows the rules set out in the International Codes[37,38]. For example, these stipulate that the naming authority be given after every scientific name. In the example above, the "L." after *Alcea rosea* stands for Linnaeus, the famous Swedish botanist of the eighteenth century. An explanation of the abbreviated names of botanical authors can be found in most comprehensive botanical or horticultural references.

The common name or names are given under the botanical name(s). Although it was not practical to include all common names, a reasonable number are listed. Many plants are also described by their French common name or names. All common names which appear in the text, English and French, are listed in the Index of Common Names.

A few of the plants listed here are named cultivars (varieties of a single species). These plants have originated and persisted under cultivation. The cultivar or variety name follows the specific name (e.g. *Melilotus alba* 'Annua' or *Melilotus alba* cv. *Annua*, indicating a special annual type of white sweet clover).

4.2 Choice of plants

All the species described here were chosen because they have or have had a favourable and established reputation among beekeepers. Commercial availability of seed or plant material was not a consideration since this is subject to change. However, except in cases where these are very important (e.g. *Taraxacum officinale*, dandelion), bee plants which would be extremely unlikely to be cultivated have not been included (e.g. *Plantago lanceolata*, plantain).

Fairly often closely related plants (i.e. closely related species belonging to the same genus) have similar value as bee plants. For instance, in general, most of the asters are comparable bee plants. However, this is not always true. *Salvia sclarea*, clary sage, is a very good bee plant, while *Salvia splendens*, often grown in gardens, is not even visited.

4.3 Arrangement of species

A horticultural arrangement has been used to group plants. They are listed in alphabetical order by scientific name in one of the following categories:
  (1) annuals and biennials
  (2) perennials
  (3) trees and shrubs

A number of lists and tables found in Section 13 are included to help the reader select plants for particular purposes or periods of bloom.

4.4 How plants are described

Each plant is described according to the arrangement noted below, with items 1 - 5 giving the basic information at a glance, and items 6 - 9 giving more detail about the plant's value as a nectar or pollen source:

Item

1. Horticultural designation: annual; biennial; perennial; tree; shrub; shrub-like tree; tree-like shrub.

2. Hardiness rating (where known): Canadian Plant Hardiness Zone (1 - 10), or typical minimum temperature range tolerated (°C). No ratings or temperatures are given for annuals.

3. Months and duration of bloom (when known).

4. Height range (m, cm).

5. Whether nectar, pollen or honeydew producer: nectar, N; pollen P; honeydew, D; ( ) indicates 'of much less importance', e.g. N(P), pollen is much less important than nectar; ? indicates uncertain information, e.g. (N?)P, pollen is collected but it is uncertain whether or not nectar is collected.

6. **Value for honey** (when known): an indication of the potential quantity of honey which could be harvested (**HP 1 - 6**)[a] ; nectar sugar concentration; honey yield per colony; nectar flow characteristics; verbal description of reputation; names of plants in same genus which are not described here but are also attractive to bees.

7. **Value for pollen** (when known): an indication of quality, quantity produced, and its attractiveness to honeybees.

8. **Honey** (when known): colour to the eye; colour class (water-white to dark amber); type of granulation; density; flavour; aroma.

9. **Notes**: brief horticultural description; cultural requirements or preferred habitat; points of general interest about the plant; range, indicating where the plant is found naturally or is naturalized in Canada and adjacent states of the United States of America; names of closely related and common native or naturalized species with unknown values for nectar or pollen are given occasionally.

In several cases, a genus (e.g. *Aster*, aster) is described in addition to individual species belonging to that genus (e.g. *Aster novae-angliae*, New England aster). This is done when there is a good deal of general information pertinent to a related group of plants. Whenever the information applies to individual species within that genus, the reader is referred back to the genus description (e.g. *Aster novae-angliae*... **Honey**: see *Aster*). This avoids repetition of the same information each time a species of the same genus is described.

4.4.1 Abbreviations used

ac.	acre
adv.	adventitious
ann.	annual
Alta.	Alberta
Apr	April
Arct.	arctic
Aug	August
auth.	authors
bienn.	biennial
B.C.	British Columbia
Calif.	California
Can.	Canada, Canadian
C.B.	Cape Breton Island, Nova Scotia
centr.	central
cm	centimetre(s)
Colo.	Colorado

---

[a] See Section 4.7 for meaning of honey potential.

Ct.	Connecticut
cv., cvs.	cultivar(s)
D	honeydew
(D?)	uncertain whether or not honeydew is collected
(D)	less important for honeydew than
Distr.	District
E.	East
e.	eastern
e.'ward	eastward
f.	filius, son or younger when following name of author
Feb	February
Fla.	Florida
Greenl.	Greenland
h	hours
H	honey
ha	hectare
HP	honey potential
I.	island
Ids.	islands
Jun	June
Jul	July
kg	kilogram(s)
km	kilometre(s)
L.	lake
Lab.	Labrador
L. Sup.	Lake Superior
m	metre(s)
Mackenz.	Mackenzie District, Canada
Man.	Manitoba
Mar	March
Mass.	Massachusetts
Me.	Maine
M.I.	Magdalen Islands, Quebec
Mich.	Michigan
Minn.	Minnesota
mm	millimetre(s)
Mont.	Montana
Mt(s)., mt(s).	mountain, mountains
N	nectar
(N?)	uncertain whether or not nectar is collected
(N)	less important for nectar than
N.	North
n.	northern
N.B.	New Brunswick
N.E.	New England
ne.	north-east(ern)
N.J.	New Jersey
N.S.	Nova Scotia
nw.	north-west(ern)
n.'ward	northward
N.Y.	New York
Nov	November
Oct	October
Ont.	Ontario
Ore.	Oregon

P	pollen
(P?)	uncertain whether or not pollen is collected
(P)	less important for pollen than
P.E.I.	Prince Edward Island
Pen.	Pennsylvania
per.	perennial
pers. comm.	personal communication with
Que.	Quebec
R.	river
Reg., reg.	Region, region
RH	relative humidity
S.	South
Sask.	Saskatchewan
se.	south-east(ern)
Sep	September
sp.	species (singular)
spp.	species (plural)
ssp.	sub-species (singular)
St. P. et Miq.	St. Pierre et Miquelon, south of Newfoundland
subarct.	subarctic
sw.	south-west(ern)
syn.	synonym
Ung.	Ungava District, Canada
USA	United States of America
var., vars.	variety, varieties
Vt.	Vermont
W.	West
w.	western
Wash.	Washington
Wisc.	Wisconsin
Wyo.	Wyoming
w.'ward	westward
Yuk.	Yukon Territory

±	plus or minus, meaning give or take
×	crossed with, the symbol for a hybrid
(:	colon inside opening bracket indicates parent species of a hybrid
(?) or ?	indicates doubt

Figures or words connected by a short dash indicate the extremes of variation (i.e. 5-10 cm means varying from 5 to 10 cm in length or height).

Figures following the abbreviation HP and separated by an oblique indicate the extremes of variation recorded for HP values (i.e. HP1/3 means that HP values of 1 and 3 have been recorded for that plant, and possibly an intermediate value as well).

References are indicated by superscript numbers (i.e. 1,2,3 etc.).

Footnotes are indicated by superscript letters in lower case (i.e. a,b,c etc.).

4.4.2 How the references are cited

Virtually all the qualitative and quantitative points describing each species are credited to a particular source or sources. Each of these sources is identified by a number from 1 to 96 (e.g. *Origanum vulgare*, wild marjoram, ... **Value for honey**: HP4[16]; N sugar concentration of 76% has been recorded, making this one of the most concentrated N's known[16];...). The sources (e.g. in this case, 16) are given in full and listed alphabetically in Section 15. In order to economize on space and limit the amount of interruption during reading, the numbers alone are used in the text.

Where no reference is given for a particular fact this is because it is common knowledge, *or* is known to be true from the author's experience, *or* it was not possible to identify the reference precisely. In many cases the original reference is given. However, in others the reader is referred to a well known source leading to one or several original references.

4.5 Plant Hardiness

Hardiness is one of the important factors to consider when selecting plants for a particular area. Many factors affect the hardiness rating of a plant, and of these the minimum temperature during the winter is the most important element in plant survival. Other important factors are length of frost-free period, summer rainfall, snow cover, and wind[81].

When possible, each plant in this book has been given a hardiness rating using the system commonly employed in Canada, Plant Hardiness Zones 1 to 10 (e.g. *Aesculus glabra*... zone 2b). This rating has been published for many cultivated trees, shrubs, climbers and ground cover plants, and takes into account the environmental factors mentioned above. It is based on the experience of horticulturalists at the Plant Research Institute at Ottawa and other stations across Canada. Some of the ratings are not final. Further testing will undoubtedly show that in some cases plants can be grown in zones colder than the ones indicated[81], and that for others the ratings are too optimistic[a]. The Map of Plant Hardiness Zones in Canada is reproduced and can be found in Section 16. Readers should identify their own zone. The rating given for a plant indicates that the plant is not likely to succeed in harsher zones.

Some of the plants described here as yet have not had their Canadian hardiness rating determined. Instead, these plants are given a minimum temperature tolerance range (e.g. *Aster laevis*... -29 to -23°C...). Minimum temperature tolerated is a less accurate measure of a plant's hardiness than the zone system, because it ignores all factors other than temperature. However, it remains a useful substitute.

A generalized map of the continental United States and Canada showing areas corresponding to these minimum tolerance ranges can be found in Section 16. Although this second map is entitled "Hardiness Zones of the United States and Canada", the ten areas shown do not correspond to the ten areas of the Map of Plant Hardiness Zones in Canada. The two maps should not be confused.

---

[a]   Brehaut, K. (1986). *Personal communication.*

4.6 Blooming dates

Blooming dates vary from year to year, depending on the earliness or the lateness of the season. And, as everyone knows, spring and summer advance across Canada from south to north. Dates of first flowering are also affected by altitude, so that the higher up a mountain a plant is grown the later it will bloom. Most of the blooming dates given in the descriptions apply fairly well to zone 6b (this includes, for instance, Halifax, Toronto, Nelson B.C.).

Whatever the *date* of blooming, the *sequence* in which a community of plants blooms is nearly always the same, and plants which bloom together will do so regardless of where they are growing. The lists of plants in order of their season of first bloom (Section 13.4) are designed to help readers plan gardens for continuous bloom. The blooming dates given should be adjusted for the reader's locality.

4.7 Honey potential

Where a plant's honey potential (HP) has been recorded it appears in the entry on a scale of 1 to 6. This potential is a measure of the quantity of honey (kg) that, theoretically, could be obtained in the course of one season from 1 ha of land growing the plant in question[16], as shown in Table 4.

To achieve this potential it is assumed that soil and weather conditions are optimal, and that sufficient foraging bees are available to collect all the nectar secreted[16]. Although these yields are seldom harvested they do give a sensible basis for comparisons between species, and are included for this reason. Honey potentials have only been estimated for a relatively small number of species.

TABLE 4. Honey Potential Classes (from Crane[16]).

Class	Honey Potential(kg/ha)[a]
1	0 - 25
2	26 - 50
3	51 - 100
4	101 - 200
5	201 - 500
6	> 500

[a]kg/ha are approximately equal to lbs/acre.

5. DESCRIPTIONS OF PLANTS

5.1 Annuals and Biennials

Alcea ficifolia L.
(syn. *Althaea ficifolia* L.)
figleaf hollyhock, Antwerp hollyhock
bienn.   Jul-Aug   1.8 m   (N)P
Value for honey and pollen: attractive to bees for N and P[35].
Notes: lemon yellow to orange flowers.

Alcea rosea L.
(*Althaea rosea* (L.) Cav.)
hollyhock, garden hollyhock, rose trémière
bienn.   Jul-Sep   1.2-2.7 m   (N)P
Value for honey: HP4[16]; N sugar concentration 34% (average)[94]; claimed to yield N freely[72], but is less attractive for N than P[35]; very attractive[79,94].
Value for pollen: P appears very attractive to bees[35]; not unusual to see 2-3 bees in a single flower collecting P[35]; abundant P, but also has been reported to be ignored by foraging bees[94].
Notes: a very popular garden plant; many colours available; sometimes persists after cultivation[23].

Borago officinalis L.
borage, cool tankard, bourrache
ann.   late Jun-frost   15-70 cm   N(P)
Value for honey: HP4[16]; N sugar concentration up to 52%, yields N all day[35]; usually not as attractive as *Trifolium repens* (white clover, trèfle blanc) or *Tilia* (basswood)[72]; continues to yield in cold weather[72]; one of the best annuals to plant for bees[50]; inverted blossom protects N from dilution by rain or dew[35]; found not to grow in competition with weeds when grown in a plot for bees[79].
Value for pollen: P collected.
Honey: probably unknown in pure form[35]; may be dark[50,72].
Notes: commonly grown for bees and as a herb; seeds freely; cultivated and persistent, often spreading to waste places; occasional, Nfld. to Ont., S. to N.S., N.E. and beyond Can. limits[23]; leaves thought to give courage and were eaten by the Crusaders for this reason.

Brassica (see individual spp. for more detail)
cabbage, canola, cauliflower, brussel sprouts, broccoli, kale, mustard, rape, rapeseed, turnip, chou, moutarde, chou broccoli, chou-fleur, chou vert
ann. bienn. per.   hardiness dependent on cv.   Jun-Oct   N(P)
Value for honey: HP2[17]; most are considered excellent N sources and yield N freely[35]; *Brassica* flowers attract bees and yield surplus where grown for seed on a large scale[35,54].
Value for pollen: varies with the spp.
Honey: light in colour; most granulates rapidly[16,17,35].

Notes: This group includes important crop plants grown chiefly in w. Canada and Ont. (i.e. canola cvs. and mustard). Use of common names without botanical names in the literature has led to confusion between the spp. to which the different cvs. belong.

Brassica napus L. ssp. *napus* [51]
(syn. *Brassica oleifera* Moench; *B. napus* L. ssp. *oleifera* Metzg.; *B. napus* L.; *B. praecox* Waldst. et Kit. ap. Schult.; *B. napus* L. var. *oleifera* (Moench) Delile; *B. campestris* L. var. *pabularia* DC.; *Napus oleifera* (Moench) Schimp. et Spenn.; *B. napus* L. var. *pabularia* (DC.) Rchb.; *B. rapa* L. ssp. *napus* (L.) Schübl. et Mart.; *B. napus* L. ssp. *pabularia* (DC.) Janchen; *B. campestris* L. var. *napus* (L.) Babingt.; *B. rapa* L. ssp. *napus* (L.) var. *arvensis* (Duch. in Lam.) Thell. f. *annua* (Schübl. et Mart.) Thell. et f. *biennis* (Schübl. et Mart.) Thell; *B. napus* ssp. *oleifera* Metzg. ap. Sinskaja; *B. napus* L. var. *pabularia* (DC.) Rchb.) [51]
Argentine type rape, rape, rapeseed, oilseed rape, swede rape, canola (certain cvs.), colza.
ann. bienn.   late Jun-Aug   0.75-1.2 m   N(P)
Value for honey: HP4/5 [16]; N sugar concentration 84% (RH 55-65%) and 23% (RH 80-90%) in outer nectaries [17]; N sugar concentration 26.8% $\pm$ 0.9% [86]; average sugar production 7.06 kg/ha/day [86]; H yield of 8 kg/colony/day reported [72].
    Fewer and larger flowers, less concentrated N, and a greater quantity of sugar produced per flower then *B. rapa* ssp. *oleifera* [86]; has been compared to *Onobrychis viciifolia* (sainfoin) and held by some to be second only to *Tilia* (basswood) [72]; 68% of the flowers of the Regent cv. stayed open for two days and 2.3% of these remained open on the third day [86].
Value for pollen: P protein is high [17]; P grains are over represented in the H [17]; honeybees collect N and none collect P only [25]; P may be discarded from body hairs or packed into P baskets [35]; produces about 1/2 the amount of P as *B. rapa* ssp. *oleifera* (i.e. about 9.33 kg/ha/24h) [86].
    Cvs. of this ssp. are self-fertile and it appears that additional bee pollination is not necessary [9,23,46,69] (in 59). However, 2 cvs. (Janetzke and Lambke) matured in a bee-free environment yielded few seeds [96] and 5 insects per flower said to be required for good pollination of this ssp. [75]; reports of increases in seed yield run from 13-95% [21,41,96] (in 59); seed yield increases attributed to insects for cvs. Altex, Andor, and Regent were 77, 49, and 58% respectively (Mohr and Jay unpubl. data) [59].
Honey: bright to pale yellow in colour; classed as white to light amber; granulation very rapid (within a few days in the comb); fine homogeneous grain; flavour sweet, mild and indistinct when granulated; aroma of the flowers can be unpleasant [17].
    H's from new canola cvs. (i.e. those cvs. low in erucic and eicosenic acid, two anti-nutritional factors) are thought to be superior in flavour to most of the winter type rapes (high in both erucic and eicosenioic acid), grown chiefly in Europe.
Notes: about 30% of all the canola grown in Canada is *B. napus* ssp. *napus*, the remainder is *B. rapa* ssp. *oleifera*. See Table 18 and *B. rapa* ssp. *oleifera*.

Brassica nigra (L.) Koch [51]
(syn. *Sinapis nigra* L.)
black mustard, moutarde noire, séneve noire
ann.   Jul-Oct   1.8 m   NP
Value for honey: H yield 27 kg/colony/season [17]; considered to be a fickle yielder [72]; H yield said to vary with climatic conditions [72]; N flow fails in some seasons [17].

Value for pollen: P protein is high[17].

Honey: light in colour with a yellow cast; classed as light amber; flavour is usually mild, but strong when freshly extracted, like mustard; aroma strong when fresh[17]; water content in ripe H may be high and fermentation reported likely[17]; H considered inferior to that from *Sinapis alba* (white mustard), having a stronger flavour and coarser grain[35].

Notes: more drought tolerant than rape; grown mostly in the prairie provinces, especially s. Alta., and Sask.; principal source of table mustard.

Brassica rapa L. ssp. *oleifera* (DC.) Metzg.[51]

(syn. *Brassica campestris* L.; *B. napus* L.; *Raphanus campestris* (L.) Cr.; *B. napella* Vill.; *B. campestris* L. var. *oleifera* DC.; *B. rapa* L.C *oleifera* DC.; *B. napus* L.A. *oleifera* DC.; *B. campestris* L. ssp. *oleifera* (DC.) Schübl. et Mart.; *B. rapa* L. convar. *oleifera* (DC.) Alef.; *B. campestris* L. ssp. *campestris* ap. Lipski.; *B. rapa* L. var. *sylvetris* (Lam.) Briggs; *B. rapa* L. ssp. *sylvestris* (Lam.) Janchen in Janchen et Wendelbg.; *B. campestris* L. ssp. *eu-campestris* (L.) Olsson; *B. rapoeuropea* Sinsk. ssp. *oleifera* (DC.) Sinsk.)[51]

Polish type rape, turnip rape, winter rape, field mustard, canola (certain cvs.), navette

ann. bienn.   Jun-Aug   50-90 cm   N

Value for honey: HP3[16]; H yields 7-14 kg/colony/season[17]; H yield said to be 40 kg/colony/season or more by beekeepers in w. Canada;[a] 9.47 kg/ha/24h[86]; N considered to be less available than from *B. napus* ssp. *napus*[61,85]; duration and intensity of the flow can differ depending on the weather conditions (also holds for *B. napus* ssp. *napus*)[86]; produces about twice as many flowers as *B. napus* ssp. *napus* (i.e. 14.21 flowers/day/plant)[86]; flowers smaller than those of *B. napus* ssp. *napus* [60]; all flowers of the Candle var. remained open for two days[86].

Considered a better H source than *Trifolium repens* (white clover, trèfle blanc) and some say that it is the best H plant known[54]. Foraging honeybees found in large numbers up to 4 km from the nearest apiary[54].

Value for pollen: not considered as important for P, but honeybees are thought to be necessary for efficient pollination[54]; produces about 20 kg/ha/24h or about twice as much as *B. napus* ssp. *napus*[86].

However, even if bees are not actively collecting P, bees serve as efficient pollinators because they almost always brush against anthers of any flower they visit[59]. Honeybees found to increase seed yield[45,58,76 (in 59)]; yield increase attributed to insects for cvs. Candle and Tobin were 139% and 81% respectively (Mohr and Jay unpubl. data noted in 59).

Honey: is white to yellowish white in colour and clear[17]; classed as light amber[17] to white[a] ; granulation rapid[17]; flavour like wine or rather like the plant[17]; or may be like *B. napus* ssp. *napus*[86]; H's from new canola cvs. (i.e. those cvs. low in erucic and eicosenic acid, two anti-nutritional factors) are thought to be superior in flavour to most of the winter type rapes (high in both erucic and eicosenioic acid), grown chiefly in Europe.

Notes: accounts for about 70% of the acreage planted to canola in Canada, remainder is *B. napus* ssp. *napus*[54]; grown for the oil contained in the seed[54]; not drought tolerant[54].

---

[a] McCutcheon, D. (1986). *Personal communication.*

DESCRIPTION OF PLANTS

Planting fields of early maturing *B. rapa* L. ssp. *oleifera* cvs. adjacent to fields of later flowering *B. napus* ssp. *napus* cvs. will prolong the season; delayed seeding of certain fields will also lengthen bloom and increase H yields.

Canola plants selected for high seed yield will also have a greater value as N and P sources since the average number of flowers per plant, seed yield, N sugar concentration and production also increase[86]. Most varieties and breeder lines grown in Canada have low erucic and eicosenioic acid (canola). Candle is one of the most popular var. and is grown extensively in the Peace River region of Alta.[86]. Canola is now becoming an important crop in Ont. See Table 18.

Campanula medium L.
Canterbury bells, violette de Marie, campanule, carillon
ann. bienn.   May-Jul   0.6-1.2 m   NP
Value for honey: beekeeper in Victoria, B.C. claimed this was more attractive to bees than catmint (*Nepeta cataria*)[72]; most *Campanula* seem to be attractive, including *C. pyramidalis* L. (chimney bellflower), *C. carpatica* Jacq. (Carpathian bellflower), and *C. persicifolia* L. (peach-leaved bellflower)[35].
Notes: culture may be difficult in some areas of Canada; lovely spikes of bell-shaped flowers, usually blue or white.

Carthamnus tinctorius L.
safflower, false saffron, bastard saffron, carthamé des teinturiers, safran bâtard
ann.   1 m   NP
Value for honey: H yield of 12 kg/colony/season reported[50]; density of 8 bees/yd^2 reported on field of safflower[72]; N sugar concentration found to be 10-15% higher than neighbouring alfalfa field[72]; yields N very freely[35]; N sugar concentration remained above 23% from 0800-1600h, and honeybees foraged in the morning, with peak activity between 0900-1100h[4].
Honey: rather dark with a good to unpleasant flavour; strong[16].
Notes: cultivated crop, grown for oil; needs summer heat to seed effectively; flower heads yield an important yellow dye.

Centaurea L. (see individual spp. for more detail)
cornflower, bluet, knapweed, thistle, star-thistle, centaurée
ann. (sometimes per.)   May-Sep   N(P)
Value for honey: HP3[16]; all are generally considered to be reliable yielders[91]; *C. nigra* (knapweed, Spanish buttons, centaurée noire) and *C. repens* (Russian knapweed) are both considered valuable to the beekeeper but are serious pests in some areas[3].
Value for pollen: minor value for P.
Honey: light amber in colour; granulation soft; thin, some with sharp flavour and characteristic after-taste[16].

Centaurea cyanus L.
common cornflower, batchelor's button, ragged sailor, blue-bottle, bluet, barbeau, bleu-bleu, casse-lunettes
ann.   Jun-Sep   0.75 m   NP
Value for honey: HP 2/3[17]; N sugar concentration 30-60% (latter is exceptional)[17]; N concentration increases during the day up to 1700h[72]; considered a good source, and bees work it all day[72].
Value for pollen: P protein value is high[17].

Honey: is greenish-yellow to dark yellowish-amber in colour; almond flavour, may be bitter and strong[17].
Notes: blue, violet, pink or white flowers, escaped from cultivation to roadsides etc.; introduced[23].

Centaurea moschata L.
sweet sultan, centaurée ambrette
ann.   half hardy   Jul-Aug   0.6 m   N
Value for honey: well liked by bees for N[35]; see *Centaurea*.
Value for pollen: see *Centaurea*.
Honey: see *Centaurea*.
Notes: sometimes spreads from cultivation[23]; showy yellow, purple or white flowers.

Cheiranthus allioni
Siberian wall-flower
bienn.   -37° to -29° C   Jun-Jul   1 m   NP[36]
Value for honey: all wall-flowers (except double) reported to be visited by bees, many constantly[35].
Notes: orange flowers in loose clusters; useful for edging; plants in cultivation known by this name or as *Erysimum allionii* may belong to *E. hieracifolium* L.. The name *Cheiranthus allioni* has no botanical standing, but is commonly used.

Chieranthus cheiri L.
(syn. *Erysimum cheiri* (L.) Cr.)
true wall-flower, common wall-flower, English wall-flower, giroflée jaune, herbe au chantre, giroflée violier
ann.   May-Jul   0.45 m   NP[36]
Value for honey: similar to *C. allioni* (Siberian wall-flower).
Notes: gold, red, and mahogany flowers; easy to grow where summers are moist and cool; requires full sun.

Clarkia unguiculata Lindl.
(syn. *Clarkia elegans* Dougl., not Poir.)
rose clarkia
ann.   hardy   Jun-Jul   0.6-0.75 m   NP[36]
Notes: lavender or rose to purple flowers; best suited to parts of coastal B.C.

Cleome hasslerana Chodat
(syn. *Cleome spinosa* of auth. not Jacq.; *C. arborea* Hort., not HBK; *C. giganteum* Hort. not L.; *C. grandis* Hort.; *Neocleome spinosa* (L.) Small)
spiny spider flower, spider flower
ann.   Aug-frost   0.3-1.0 m   NP
Value for honey: secretes N abundantly in the early morning and again just at nightfall[72]; known to be one of the very best bee plants when grown under favourable conditions[72]; needs rich soil to yield well[72].
Notes: spikes of pink and white fragrant flowers; rather coarse and needs room to develop; sometimes runs wild in the e. USA; similar to and is confused with *C. spinosa* Jacq., spiny spider flower.

Cleome lutea Hook.

yellow spider plant, golden cleome, yellow bee plant

ann.   0.6 - 1.0 m   NP

Value for honey: N sugar concentration 23% (average) also 12% in another area[79]; N abundant[20]; a tenth acre plot encouraged later brood rearing in an adjacent apiary, compared to other colonies without this source[72]; in some areas is more important for P[67]; may have been principal host for alkali bee before such plants as alfalfa and sweet clover were introduced[67].

Value for pollen: a good source of P[72]; P supplied in the morning[79].

Honey: dark and strong[79].

Notes: considered weedy.

Cleome serrulata Pursh

(syn. *Cleome integrifolia* (Nutt.) Torr. & A. Gray)

bee spider flower, Rocky Mountain bee plant, stinking clover

ann.   Jul-Sep   0.6-1.0 m   NP

Value for honey: yield of 46 kg/colony/season reported (a 10-day flow)[72]; 2-3 super surplus per colony in 3 weeks[50]; N sugar concentration lower than alfalfa or sweet clover and is probably not preferred to these[90]; more important in the W., probably a source of surplus in s. Sask., and coastal B.C.[72] (a beekeeper in Victoria, B.C. claims that surplus from this sp. is unlikely in coastal B.C.[a]); also said to yield N heavily under cool cloudy conditions that stops clover from yielding[72]; yields earlier in the day than sweet clover[72]; beekeepers on the Eastern Seaboard tried to introduce this by freely sowing seed in waste land[79] (early in century).

Value for pollen: bees hover humming-bird fashion while collecting P[79].

Honey: white with greenish tinge[50]; flavour variable, poor at first[56], but improves with age[72].

Notes: prefers moist soil; unpalatable to livestock[72]; minor ornamental interest; dense showy clusters of white flowers on each plant; prairies, damp soils and waste land, Man. to Wash. and S. beyond Can. limits[23].

Convolvulus tricolor L.

dwarf morning glory, heavenly blue, dwarf glorybind, belle de jour, liseron tricolore

ann.   Jul-Sep   0.2 m   NP[36]

Value for honey: H yield of 32 kg/colony/season reported from wild morning glory (*Calystegia sepium* (L.) R. Br., syn. *Convolvulus sepium* L.)[72]; *C. arvensis* L. (field bindweed) is a source of wild bindweed (morning glory) H, but is also a troublesome weed in some areas[3].

Honey: white and good flavour[72].

Notes: must remove dead flowers regularly to prolong bloom.

Coreopsis L. spp.

coreopsis, tickseed, coréopsis

ann.   Jul-Aug   15 cm   NP

Value for pollen: attractive sources for honeybees and many of the spp. (as opposed to the garden forms) are freely visited[35]; a useful late source.

Notes: often confused with *Bidens* (Spanish needle) which is also a good source of late forage; choose singles only; large group of herbs, grows well in any garden soil.

---

[a]   Mitchell, S. (1987). *Personal communication.*

Coriandrum sativum L.

coriander, Chinese parsley, coriandre

ann.   Jun-Jul   1 m   NP

Value for honey: HP5[16]; low N sugar value[17] (i.e. weight of sugar produced in a 24h period); fairly long bloom of 1 month[17]; bees visit freely[72].

Honey: is dark in colour[17].

Notes: cultivated crop or garden plant grown for leaves, seeds and oils; found in waste areas, chiefly spread from cultivation[23].

Cosmos bipinnatus Cav.

cosmos, cosmos bipinné

ann.   Aug-frost   3 m   N[36]

Value for honey: attracts bees in large numbers, and is a useful late source[72].

Notes: obtain early flowering cvs. because long growth season is required and late flowering cvs. may not have enough time to bloom in Canada; soil should not be too rich; roadsides and waste land, spread from cultivation; becoming weedy and more common beyond Can. limits[23].

Digitalis purpurea L.

foxglove, common foxglove, digitale pourprée

bienn. (sometimes per.)   -28°to -23°C   Jun-Jul   60-120 cm   NP

Value for honey: a H flow of 5 weeks resulted in an average of 4 filled supers per colony[72]; generally more attractive to bumblebees than honeybees[79]; said to make a significant contribution to H surplus in parts of B.C.[72]; bees able to secure a full load from one flower[35].

Value for pollen: where foxglove is extensively cultivated, its P has poisoned bees[35].

Honey: golden yellow in colour; flavour is good[72].

Notes: popular garden plant; numerous purplish flowers on a long spike that open progressively.

Dipsacus sativus (L.) Honckeny[51]

(syn. *D. fullonum* var. *sativus* L.; *D. fullonum* sensu Hudson et auct. americ.; *D. fullonum* L. ssp. *sativus* Thell.)[51]

fuller's teasel, teasel, cardère, chardon à foulon, chardon à bonnetier

bienn.   late Jun-Sep   50-200 cm   NP

Value for honey: N sugar concentration 26%[17]; has been described as one of the best H plants in existence[72]; an important source of surplus where commercially grown[36]; begins yielding at the same time as *Tilia* (basswood) or *Trifolium* (true clover); bees visit this all day; *D. lacinatus*, sometimes grown ornamentally, is noted to be attractive[35]; *D. sylvestris* (wild teasel) is apparently equally attractive[35].

Notes: grown ornamentally, chiefly for use in dried flower arrangements; formerly cultivated for heads of stiff bristles used to raise the nap on woollen cloth (e.g. in N.Y.); still cultivated for the latter purpose in the U.K.[35]; *D. sylvestris* is found in roadsides, old fields and pastures, w. Que. and w. N.E. to Ont.[23].

Dracocephalum moldavica L.

annual dragon's head, Moldavian balm, dracocéphale Moldavique

ann.   Jul-late Aug   45 cm   N

Value for honey: HP5[16]; N sugar concentration 30-40%[16]; said to attract bees consistently all day[35].

Honey: light in colour; tastes slightly of lemon[17].
Notes: traditionally used in Europe for rubbing inside of straw skeps to attract swarms[35]; some shade, moist soil; waste or cultivated ground in local areas N.E. to Wisc. and beyond Can. limits[23].

Echium vulgare L.
viper's bugloss, blue-weed, blue thistle, blue devil, viperine, langue d'oie
bienn.   -37° to -28° C   Jun-Sep   20-80 cm   NP
Value for honey: HP6[16]; N sugar concentration 17-43%, maximum at 1500h[17]; an important source where it is common[78] (e.g. common in s. Ont.[3]); *E. lycopsis* or *E. plantagineum* (called salvation Jane by beekeepers in Australia) is an important H source in Australia as well as an invasive weed in pastures.
Honey: white to light amber in colour[16]; dull in appearance when granulated, with a fine grain[91]; delicate but flat flavour.
Notes: considered an invasive weed in some areas[3]; seeds easily; in some countries this is sown on railway embankments and road verges to increase bee forage (e.g. in Germany)[35]; roadsides, dry fields and waste land, often an obnoxious weed, Que. to w. Ont., S. to beyond Can. limits[23]; introduced; see Table 18.

Eschscholzia californica Cham.
California poppy, poppy, eschscholtzie de Californie, pavot de Californie
ann.   Jul-frost   60 cm   (N)P[36]
Value for honey: probably only yields N erratically or in small amounts[35].
Value for pollen: abundant P, relished by bees[35]; valuable fall source.
Notes: these do well along coast as they tolerate salt spray; widely distributed in Calif.

Fagopyrum esculentum Moench[51]
(syn. *Fagopyrum vulgare* Hill; *Polygonum fagopyrum* L.; *P. tataricum* Lour.; *P. saggittatum* Gilib.)[51]
buckwheat, notch seeded buckwheat, sarrasin, blé noir
ann.   Jul-Aug   1 m   N(P)
Value for honey: HP3/5[16]; H yield is variable, but if conditions are favourable, a strong colony will store 3-4 kg/ha/day and 25 kg/ha may be stored[72]; 60 kg/ha is stored in an exceptional season[35]; some beekeepers claim that 30 colonies need 1.6 ha[72]; N flow is copious for 2-3 weeks, but 70% of the sugar is secreted in the first half of flowering[17]; N flow is usually between 0900-1400h[72]; N sugar concentration varies from 7-48%[17]; bees may become "angry" in the afternoon when the flow ceases[36]; N flow and quality are strongly affected by environmental conditions (lime, nitrogen and phosphorus in the soil improves sugar concentration by 20-50%)[17].

Cool moist conditions at flowering are best, and may allow bees to work blossoms all day[35]. Hot dry weather may terminate flow[35]. On fertile soil with adequate moisture, secretion is maintained at low temperatures[17]. Cool nights and mean temperatures of less than 21°C are best (optimum range is 16-26°C)[17]. N secretion is increased by dusting crop with boric acid[17].

Does not yield well on heavy soil (e.g. known not to yield well in the St. Lawrence Valley)[72]; better reputation as a bee plant in Ont., Que., and parts of the Maritimes than in the W.[72].

When planting for bees, 3-4 successional sowings are recommended for a 2-3 month bloom period. In order to initially attract bees to the crop, a small percentage of an earlier flowering cv. could be planted[35].

This is one of the few plants that offers the possibility of being sown as artificial bee pasturage on an economic basis since the cultivation costs are offset by the resulting crop[35].

Value for pollen: P protein value is high, but dried P may be toxic to bees[83]; P yield is low to absent[17]; over-represented[a] in H[17].

Honey: opinions vary, but most observers agree that it is dark brown to reddish or black in colour[17]; classed as dark amber; granulation is usually slow and may be coarse[17]; flavour is agreed to be strong and may seem unpleasant, bitter or burnt to some tastes[17]; aroma is strong and may be unpleasant[17]; a specialty H, in demand by particular markets and sells well above premium prices[78]; usually in short supply[78].

Notes: often grown on low fertility soil that is being brought back into cultivation; useful for protecting roots of saplings, for green manure, and game coverts[35]; not suitable on heavy clays (sandy soil best); very acid soil is tolerated; very few insect or disease problems; see Table 18.

Fagopyrum tataricum (L.) Gaertn.[51]
(syn. *Polygonum tataricum* L.; *Fagopyrum tataricum* var. *vulgare* Alef.)[51]
Tartary buckwheat, Indian buckwheat, buckwheat, Kangra buckwheat, sarrasin de Tartarie
ann.   Jul-Aug   60 cm   N(P)
Value for honey and pollen: similar to *F. esculentum* (buckwheat). and is usually grown in e. Canada; may be considered a weed in w. provinces.

Gaillardia pulchella Foug.
painted gaillardia, Indian blanket flower, blanket flower
ann.   Jun-Oct   30-60 cm   NP
Value for honey: H yield of 34 kg/colony/season reported where it grows widely (e.g. Texas)[50]; N sugar concentration 32%; honeybees work flowers vigorously[94].
Value for pollen: P is also collected[35,94].
Honey: yellow amber in colour; flavour fair and considered of low quality[50].
Notes: entirely yellow, daisy-like flowers, or yellow with red tips; also red, tipped with yellow or entirely red; full sun preferred; dry, sandy soil, prairies and openings in W.; casually as a migrant or garden escape along Atlantic coast[23].

Gilia capitata Sims
globe gilia, gilie
ann.   Jul-Sep   45 cm   NP[36]
Notes: blue, violet or white flowers, 50-100 in terminal heads about 3 cm across; native to B.C.

Helianthus L.
sunflower, cornflower, tournesol
ann. (per.)   Jul-Oct   N(P)
Value of honey: most of these are considered valuable N plants.
Honey: usually dark and strong.

---

[a] This means that the percentage of pollen grains of this species to the total pollen grains in honey will be characteristically high even when most of the honey surplus does not originate from this species.

Helianthus annuus L.
common sunflower, grand soleil, fleure de soleil, tournesol
ann.   Jul-Aug   1-3 m   N(D)
Value for honey: HP2[16]; H yield usually 30 kg/colony/season[17]; N sugar concentration 50%, up to 60% (in hot areas)[17]; N sugar concentration and volume peaks reached together[52]; N secretion highest during first 10 days of bloom, and in plants with long daily exposure to light[17]; one of the few plants said to be worth growing for H production[90]; extra-floral nectaries are also visited.
Honey: yellow or golden in colour; classed as amber; granulates rapidly to a fine soft grain; flavour is mild, characteristic; aroma fairly strong[17] and unpleasant to some[72].
Notes: gummy secretion from flower heads may be collected, probably as propolis[35]; deep moisture-retentive and nitrogen-rich soil preferred[17]; huge copper, yellow or bronze flowers, each having 100-2000 florets; found throughout s. Canada, especially bottomlands and other rich soil in Man. and beyond Can. limits; cultivated and spread to waste land etc. E. to Que., N.B., N.S.[23].

Helianthus petiolaris Nutt.
Prairie sunflower
ann.   late Jul-early Sep   180 cm   N
Value for honey: N sugar concentration 49%; in certain areas bees observed to visit this in preference to *Melilotus officinalis* (yellow sweet clover)[94]; plants growing on dry sites were neglected by bees, while those in moist soil in the same field were attractive to bees[90].
Honey: probably similar to *H. annuus* (common sunflower).
Notes: sometimes grown as an ornamental; naturalized from Sask. W. in sandy soil, dry roadsides, along railroads and in waste land, Man.[23] and doubtless other areas.

Heliotropium curassavicum L.
seaside heliotrope, héliotrope, héliotrope de Curçao
ann.   Jul-Sep   15 cm   creeping   N
Value for honey: very attractive to bees[35]; this is an important source of N where it is common[72].
Notes: creeping and forms a mat; small white or blue flowers with a yellow eye; withstands seaside conditions; useful ground cover, especially in alkaline or saline areas.

Iberis umbellata L.
globe candy-tuft, thlapsi des jardiniers, téraspic d'été
ann.   Jul-Sep   40 cm   NP[36]
Notes: sow at intervals and remove pods for continuous and longer bloom; flowers appear in compact clusters in several colours; not fragrant.

Impatiens glandulifera Royle
Himalayan balsam, touch me not, impatiente, n'y-touchez pas
ann.   Aug-Sep   75 cm   NP
Value for honey: Honeybees disappear out of sight in these 30 mm long blossoms when collecting N in the long curved spurs at the bases of the flowers. Most of the *I. balsamina* (garden balsam, balsamine) cvs. are doubles and offer nothing to the bee[35].
Notes: sun or shade; coarse appearance; interesting fruits which burst suddenly and expel the seeds; fleshy stems; showy blooms; waste land and roadside thickets, N.S., n. N.B., n. N.E., Que., and Ont. where escaped from cultivation[23].

Isatis tinctoria L.

dyer's woad, asp-of-Jerusalem, pastel des teinturiers, pastel cultivé, herbe de Saint Phillipe

bienn.   Jun-Jul   1.2 m   N

Value for honey: HP[16]; attractive for N[35].

Notes: no longer commercially grown for blue dye obtained from its leaves; numerous bright yellow flowers in terminal panicles; can be grown as an edging plant; prefers alkaline soil; roadsides, in local areas, se. Nfld. and doubtless elsewhere[23].

Lavatera L. (see individual spp. for more detail)

mallow, tree mallow, mauve

ann. bienn. per.   Jul-Sep   (N)P

Value for honey: most yield at least some N.

Value for pollen: abundant P[35].

Notes: a large group of ornamentals; most grow very tall in one season and so can be used for temporary screening or as background in border; full sun.

Lavatera trimestris L.

(syn. *Lavatera alba* Medic.; *L. rosea* Medic.)

herb tree mallow, lavatere àgrandes fleures, mauve royale

ann.   Jul-Sep   180 cm   NP

Value for honey: HP3[16]; this mallow is unusual in being very attractive for N.

Value for pollen: abundant P.

Notes: useful for back of the border because of extreme height; white or rose to red flowers are funnel shaped and 10 cm across; sow where plants are to stand.

Leonurus sibericus L.

chivirico, motherwort, agripaume

ann.   N

Value for honey: a very good to excellent N plant; covered with bees until frost according to Pellet Gardens' catalogue, 1978.

Notes: reseeds freely, but not particularly difficult to control; gray-green pubescent leaves; waste land, locally N. to Que. from s. areas, beyond Can. limits[23].

Limnanthes douglasii R. Br.

meadow foam, Douglas meadow foam, limanthes

ann.   Jun   15 cm   NP

Value for honey: very attractive source of N; N secreted abundantly[35].

Honey: is light in colour, with a good flavour[35].

Notes: profusely blooming; sunny locations and moist soil preferred; important to grow in mass for maximum attraction; native to parts of coastal B.C. and, when blooming, makes a striking sheet of colour on low ground.

Limonium sinuatum (L.) Mill.

(syn. *Statice sinuatum* L.)

statice, notch leaf sea lavender, statice sinué

bienn.   Aug   60 cm   N[35]

Value for honey: probably similar to *L. latifolium.*

Notes: long branching sprays of small purple flowers.

Linum usitatissimum L.

flax, annual flax, linseed, lin

ann.   Jun-Jul   40-80 cm   NP

Value for honey: HP1[16]; N sugar concentration of 26-49%[16]; in Brandon Man. bees were reported foraging on a 40 ha flax field while sweet clover (*Melilotus*) was blooming nearby[72]; petals tend to drop in the afternoon, so bees forage in the morning only[35]; usually agreed that it has feeble attraction for honeybees.

Value for pollen: cross pollination by bees is necessary and increases seed weight and total yield[72]; P load is blue.

Notes: commercially cultivated for its seed, the source of linseed oil, linseed cake (for stock feed) and flax fiber (for the linen industry); attractive sky-blue flowers; escaped from cultivation in much of N. America.

Lobularia maritima (L.) Desv.

(syn. *Alyssum maritimum* (L.) Lam.)

sweet alyssum, alysse maritime, alysse adorante

per. (usually grown as an ann. in Canada)   Jun-frost   15 cm   N[35]

Value for honey: favoured by bees for N[35]; very long blooming season.

Notes: honey scented flowers; popular edging plant and may self seed; shearing in summer may assist in producing good late bloom; often cultivated, but occasionally spontaneous, N.E.[23].

Medicago lupulinas L.

black medick, hop clover, yellow trefoil, none-such, lupuline minette, luzerne, lupuline

ann. bienn.   hardiness dependent on cv.   Jun-Sep   NP

Value for honey: HP1/2[16]; freely visited for N (e.g. in the UK.)[35]; usually blooms earlier than white or red clover; said to be of little value as bee forage where naturalized (parts of the Pacific Northwest, such as Ore. and s. B.C.)[79]; may withstand drought and continue yielding[72].

Notes: may be suitable in a seed mix for roadsides or waste land, especially alkaline soils; a fodder plant.

Melilotus Mill. (see individual spp. for more detail)

sweet clover, melilot, vieux garçons

ann. bienn.   hardiness dependent on cv.   Jun-Sep   NP

Value for honey: HP4/6[16]; both *M. alba* (white sweet clover) and *M. officinalis* (yellow sweet clover) are major H sources in w. Canada and were also important to beekeepers in many other areas 50 years ago (e.g. Ont. and Man.)[16,57,88]; reduction of sweet clover acreage in Ont. partly explains the decline in H production over the same period[16,88]. Crops such as alfalfa have replaced the acreage, but have not compensated for the loss of sweet clover[16,57].

N yield appears best on dry as compared to wet soil[67,72]; N flow reported not to be uniform, and once stopped, does not usually begin again[79]; N producing capability of the crop can be extended by cutting back early growth to delay onset of flowering.

Total period of sweet clover bloom can be extended if early and late cvs. are grown together (e.g. *M. alba* 'Grundy', early, and *M. alba* 'Polara', late). Also one or both of the annual forms might be included to provide bloom the first year (e.g. *M. alba* 'Annua' and *M. indica*).

Value for pollen: generally less important than N.

Honey: white or greenish-yellow in colour[16]; flavour delicate, like cinnamon or vanilla[16]; excellent[35].

Notes: reasons for its decline in popularity as an agricultural crop include: a tendency to persist in fields due to its hard seed coat; susceptibility to sweet clover root rot and weevil.

Sweet clover often thrives where few other plants will grow, such as banks, quarries and waste land. However, some highway departments are discouraged from planting sweet clover on road verges for ground cover because of adverse public opinion: the volunteer plants have a tendency to become a nuisance in adjacent farms, especially where seed crops are grown, or the plants are considered too tall to be safe for use along road verges (e.g. parts of B.C.).

In its earlier and more popular era, sweet clover was often recommended for naturalizing to increase bee forage. In 1938, in B.C., it was recommended that a liberal liming before sowing along a roadside would help to establish a stand of sweet clover for bee forage[73].

Useful for soils too wet or too dry for either *Medicago* (alfalfa) or *Trifolium pratense* (red clover); does best in well-drained clay or clay loam, and not in water-logged or acid soil; grows up to 2 250 m[17].

None of the sweet clovers makes a particularly attractive garden plant, but they may be suitable for some roadsides, banks or waste land to reduce soil erosion and eradicate obnoxious weeds by crowding them out. See also Table 18.

Melilotus alba Desr.
white sweet clover, bee clover, white melilot, melilot blanc
bienn.   hardiness depends on cv.   mid Jul-late Aug   50-150 cm   N(P)
Value for honey: HP5/6[16]; N sugar concentration 33% on wet soil and 55% on dry soil[67]; N sugar concentration of *M. alba* was 35% and *M. officinalis* (yellow sweet clover) was 52% on the same day and same situation[79] (some researchers have found much less of a difference)[17]; despite its lower sugar concentration, *M. alba* has a better reputation among beekeepers than *M. officinalis*[67]; this is probably due to the fact that the overall volume of N secreted by the white is probably greater than that from the yellow[16,67].

N flow starts about two weeks after the yellow (or two weeks after the first alfalfa flow) and is a little shorter than the flow from the yellow[35,79]; N flow may last for up to 7 weeks, but each flower blooms only for about 3-4 days[16].

Value for pollen: little P gathered; honeybees observed to take longer to obtain a load of P from this than from *M. officinalis*, and the loads are always small[67,79,94]; P load is yellowish-green to green or brown[17].

Honey: may have a greenish tinge; classed as white to water-white; granulates within one week of removing from the hive and forms a fine mass; heavy body; tastes of cinnamon and is mild to peppery[16].

Notes: several cvs.; naturalized in N. America; see *Melilotus*.

Melilotus alba 'Annua' H.S. Coe
Hubam clover
ann.   late Aug-early Oct   N(P)
Value for honey: H yield of 18-22 kg/colony/season reported in warm autumns[94]; N sugar concentration 30%[94]; bees said to be very active on this sp., even in cool weather[94]; see *Melilotus*.

Value for pollen: little P collected[94].

Notes: matures in one season; useful in order to extend the sweet clover bloom and usually grown with the biennial sweet clovers; can be grown for late pasturage or green manuring; waste or cultivated ground throughout most of Canada[23]; see *Melilotus*.

Melilotus officinalis (L.) Pall.
yellow sweet clover, yellow melilot, trèfle d'odeur jaune
bienn.   hardiness dependent on cv.   Jun-Sep   40-250 cm   NP
Value for honey: HP4[16]; N sugar concentration up to 52% and usually higher than that of *M. alba* (white sweet clover)[67]; less popular with beekeepers than *M. alba* (white sweet clover)[67], possibly because N is probably less abundant than from *M. alba*[67]; N secretion is reduced by insufficient rain or soil moisture[17]; N flow starts about 2 weeks before that of the white (*M. alba*); see *Melilotus* and *M. alba*.
Value for pollen: P yield is up to 5 times as abundant as from *M. alba*, and bees obtain a P load more quickly than from the latter[79]; P load is dark yellow[17].
Honey: classed as white to amber; granulates rapidly and finely; flavour mild to slightly peppery; aroma is delicate, like vanilla[17].
Notes: slightly more drought tolerant than *M. alba*; several cvs. available; waste land or cultivated ground, Que. to B.C. and doubtless many other areas[23]; see *Melilotus*.

Nemophila menziesii Hook. & Arn.
baby blue eyes
ann.   Jul   75 cm   N[36]
Value for honey: described as a favourite for N[35].
Notes: successional sowings will result in a longer bloom period; excellent for edging and massing.

Nicotiana alata Link & Otto
(syn. *N. affinia* T. Moore)
winged tobacco, flowering tobacco, jasmine tobacco
ann.   Jul-Sep   45-120 cm   NP[36]
Value for honey: probably similar to *N. tabacum* (tobacco)
Notes: large fragrant flowers in many shades of colour; rich, well-drained soil.

Nicotiana tabacum L.
tobacco, tabac
ann.   Aug-frost   180 cm   N(P?)
Value for honey: HP2[16]; bees described as "swarming" on the blossoms[72]; when commercially grown is usually cut before optimum bloom is reached, but in the early 1900's, crops were allowed to flower and a H surplus often obtained[72]; bees tend to puncture flower bases to collect N[35].
Honey: dark; strongly flavoured; heavy body, and slow to granulate[72].
Notes: when grown as a garden ann., seedlings should not be set out until all danger of frost is past; escapes from cultivation, but is rarely persistent[23].

Ocimum basilicum L.
sweet basil, common basil, basilic
ann.   Aug   60 cm   N
Value for honey: HP3[16].
Honey: very light; fine flavour[16].
Notes: culinary herb.

Origanum majorana L.

(syn. *Majorana hortensis* Moench; *M. vulgaris* Mill.; M. majorana (L.) H. Karst.)[51]

sweet marjoram, knotted marjoram, annual marjoram, marjolaine

tender per. (usually grown as an ann. in Canada)   Jul-Sep   60 cm   N

Value for honey: eagerly sought after by bees[35]; N secreted in the mornings only[52]; probably similar to *O. vulgare* (wild marjoram).

Value for pollen: offers little or no P.

Honey: may be similar to *O. vulgare*, wild marjoram.

Notes: culinary herb; well-drained soil; sun; roadsides, old fields, thin woods, sw. Que. and s. Ont.; naturalized[23].

Papaver rhoeas L.

corn poppy, field poppy, Flander's poppy, coquelicot, pavot coquelicot

ann.   Jun-Aug   1 m   (N)P

Value for honey: a certain amount of N collected[35].

Value for pollen: abundant dark P[35]; exceptionally attractive source which may have a narcotic effect on foragers and returning bees may have difficulty in finding the hive entrance[35,72].

Notes: single cvs. only; rubbish heaps and rarely old fields, N.S. and S. beyond our limits[23]; the common field poppy of Europe, Shirley poppy, is a strain of this sp.

Phacelia tanacetifolia Benth.

phacelia, fiddleneck, tansy phacelia, valley vervain, phacelie à feuilles de tanaisie

ann.   5° of frost tolerated   Jun-Aug   40-70 cm   NP

Value for honey: HP5/6[17]; H yield 5-9 kg/colony/season[17]; N sugar concentration 16-52% (latter in a dry summer)[17]; N is secreted at 10-31° C, best at 16-24° C, RH 55-70% and is increased by application of complete N-P-K fertilizer and late spring rain[17]; reported to be "covered with bees from morning to night"[72]; grown for bee forage in some countries (e.g. Germany)[35]; *P. viscida* (Benth. ex Lindl.) Torr. (sticky phacelia) and *P. campanularia* A. Gray (California bluebell) are also ornamentals (not listed) and attractive to bees[35].

Value for pollen: abundant P[17]; P load is dark blue to dull brown[17].

Honey: light green; granulates rapidly[16]; classed as white to light amber[17]; flavour fine.

Notes: blooms in 6-8 weeks from seed, and lasts 4-6 weeks in flower[72] (long blooming period for an ann.); could be sown with sweet clover biennials to provide N flow in the same season[35]; has sometimes been grown as a companion crop to potatoes, between the rows after the last earthing up, in parts of Europe (e.g. Germany)[35]; can also be dug in as green manure in the autumn[35]; nutritive value for livestock is said to be between red clover (*Trifolium pratense*) and crimson clover (*Trifolium incarnatum*), and to be more palatable than alfalfa[72]; needs full sun.

Phaseolus L. (see individual spp. for more detail)

bean, kidney bean, haricot

ann.   Jul(?)   N(P)

Value for honey: N flow from beans is regarded as being very dependable[72]; *P. lunatus* (lima bean or butter bean) and to a lesser extent, *Vigna aureus* (mung bean), are the only two beans of significant value to beekeepers in N. America[54].

Honey: light in colour; granulates rapidly; mild, undistinguished flavour[16].

Notes: many spp. are grown as food crops.

Phaseolus coccineus L.
runner bean, scarlet runner bean, haricot d'Espagne, haricot écarlat
ann.   Jul   3-4 m   N(P?)
Value for honey: surplus occasionally reported[54]; flowers are too large for the honeybee to negotiate, except between 0800-1000h, when N is high in the flower tube[35]; if N is lower in the tube, then it is only available to honeybees through "robbing" holes made in the base of the flower by bumblebees[35]; flower size may be reduced if plants are pinched back during early growth[35]; see *Phaseolus* .

Pimpinella anisum L.
anise, common anise, aniseed, anis vert
ann.   60 cm   N(P?)
Value for honey: very attractive to honeybees for N[72].
Honey: light in colour; mild, elusive flavour[72].
Notes: copious yellowish-white flowers that appear about 3 months after planting[54]; a popular garden ornamental, sometimes grown commercially in N. America for medicinal or culinary purposes; has escaped from cultivation[23].

Reseda odorata L.
mignonette, common mignonette, réséda odorante, herbe d'amour
ann.   Jun-frost   N(P)
Value for honey: sometimes considered an excellent plant for N[35]; some say that it is capable of giving more N and more blossom for a given area than any other plant known[35,72]; also has been known to have a poor attraction which may be due to the possibility that it only yields well in warm weather[50]; reported to be attractive to honeybees in Kootenay, B.C.[50]; many extravagant claims made about its value which may deserve field trials[72].
Value for pollen: little P collected[35].
Notes: flowers fragrant; cool moist site preferred; cultivation for essential oil used in perfumery.

Rudbeckia L. (see individual spp. for more detail)
coneflower, rudbeckie
Value for honey: sometimes a fair source of late N, but opinions differ[50]; occasionally reported to be unattractive to honeybees[72]; bees rarely seen on these plants in the author's garden.
Notes: a large group of wild and cultivated plants; thrive in almost any soil or location.

Rudbeckia hirta L.
black-eyed Susan, rudbeckie hérissée
ann. bienn. (or short lived per.)   Aug   30-90 cm   N
Value for honey: see *Rudbeckia* L.
Notes: well known daisy-like flowers of open woods, barrens, and roadsides suitable also for ornamental planting; see *Rudbeckia*.

Salvia sclarea L.
clary sage, clary, sclarée, toute-bonne, sauvage sclarée
bienn.   Jul-Aug   60-90 cm   N
Value for honey: HP4[16].
Honey: (for *Salvia* L.) water-white; slow to granulate; many with mild characteristic

flavour[16].
Notes: self-seeds freely; source of aromatic oil used in medicine and also for flavouring; note that *S. splendens* (scarlet sage) is not visited by honeybees.

Satureja hortensis L.
summer savory, savory, sarriette
ann.   Jul-Oct   45 cm   N
Value for honey: HP3 (for *Satureja* L.)[16]; N sugar concentration may be very high[16]; less attractive than *S. montana* (winter savory)[35]
Honey: similar to *Thymus* L. (thyme) H[16]; a major source of the famous Greek H from Mt. Hymettus[16].
Notes: fragrant pale mauve flowers in whorls or spikes; culinary herb; roadsides and waste land where escaped from cultivation[23].

Scabiosa L. (see individual spp. for more detail)
scabious, pincushion flower, scabieuse
ann. bienn. per.   Jul-Sep   N
Value for honey: most of these seem attractive to honeybees[35].
Notes: popular for flower gardens and thrives in any good soil in a sunny location.

Scabiosa atropurpurea L.
sweet scabious, mourning brides, pincushions, scabieuse des jardins
ann.   Jun-Sep   45-60 cm   N
Value for honey: bees work freely for N, especially when grown in mass[16].
Notes: needs well-drained soil; sun; naturalized in Calif.[23]; see *Scabiosa*.

Senecio elegans L.
(syn. *Jacobaea elegans* (L.) Moench)
purple ragwort, seneçon
ann.   Jun-Aug   30-60 cm   NP
Notes: sow at intervals; flowers are yellow with purple or white; escaped cultivation in Calif.; a near relation, *S. jacobaea* (stinking Willie, tansy ragwort), is a serious weed in many areas of Canada and also has a reputation for producing an undesirable table H (see Section 8).

Sinapis alba L. ssp. *alba*[51]
(syn. *S. alba* L.; *Brassica alba* (L.) Rabenhorst)[51]
white mustard, moutarde blanche
ann.   Jul-Oct   80 cm   NP
Value for honey: HP1/3[16]; N sugar concentration up to 60% in dry seasons[17]; has been recommended for planting to increase H production in some countries (e.g. USSR.)[17].
Honey: colour and some characteristics similar to *Brassica* H, especially *Brassica napus* ssp. *napus*.
Notes: cultivated for seeds, green manure and forage; waste land and roadsides in local areas, P.E.I., to B.C., S. to N.S., N.E. and beyond Can. limits[23].

Tagetes L.
marigold, oiellet d'Inde, tagète
ann.   Jun-Oct   (N?)P
Value for honey and pollen: many of the single cvs. are freely visited, especially in the autumn and when grown in mass.
Notes: very strongly scented pot herbs; popular for cut flowers and simple to cultivate.

Trifolium incarnatum L.
crimson clover, carnation clover, scarlet clover, Italian clover, trèfle incarnat, farouche
ann. bienn.   May-Jul   90 cm   NP
Value for honey: HP1[17]; H yield up to 95 kg/colony/season[17]; N sugar concentration 31-60%[17]; N secreted in the mornings only[54]; blooms earlier than other clovers, and has been ranked with buckwheat in value[72]; bees may have difficulty in collecting the N because of the structure of the flower[54]; some beekeepers notice that bees working this crop tend to swarm excessively, although they do not crowd the brood-nest with H[54].
Value for pollen: P is collected in large amounts which is unusual for a legume, but in the afternoons only[54].
Honey: yellowish in colour, or pale; classed as water-white to extra light amber; granulates to a solid paste, but grains redissolve readily; flavour delicate, and aroma varies[54]; considered excellent[54].
Notes: tolerates acid soil better than *T. pratense* (red clover) or *T. repens* (white clover); not completely hardy; used as forage and in crop rotation; occasionally spread to waste ground, roadsides etc.[23]; see *Trifolium*.

Verbascum blatteria L.
moth mullein, molène aux teignes
bienn.   -37° to -29° C   Jun-Sep   90-120 cm   N(P)
Value for honey: offers N sparingly[35].
Value for pollen: very attractive for P, but chiefly from Aug-Sep[16].
Notes: common roadside weed, but is grown in "wild gardens"; *V. thapsus* (flannel mullein, tabac du diable, bouillon blanc) is seldom cultivated, but P is usually highly attractive to bees[35]; roadsides and old fields, N.E. to Ont., S. and W. beyond Can. range[23]; naturalized (from Europe).

Vicia faba L.
broad bean, faba bean, field bean, horse bean, magazan bean, Windsor bean, fève de marais, féverole
ann. bienn.   very hardy   Jul-Aug(?)   less than 2 m   N(D)[17]P
Value for honey: HP3[16]; N sugar concentration 28%[17]; N secretion is increased by higher planting density[54]; N is also produced in extra-floral nectaries on undersides of leaf bases, and is collected from 1400-1600h when P is available as well[54]; N may be less attractive to honeybees than the P[54]; flower size varies with the variety, and the larger flowers may only be worked by honeybees using holes made by bumblebees in the base of blossoms[54] (broad bean cvs. of the vegetable garden usually have large flowers[35]).
Value for pollen: may be more attractive for P than N[54]; P load is gray-green[17].
Honey: is light in colour, but darker if honeydew is present; classed as white to dark amber; granulation is often rapid, and the grains coarse; flavour is pleasant and mild[17].
Notes: grown on a field scale for fodder or for human consumption (in fresh or dried

form); is a legume and so fixes nitrogen; grows well on heavy soil; this is the bean of antiquity; grown across Canada, occasionally escapes from cultivation[23].

Vicia pannonica Crantz
Hungarian vetch, vesce de Pannonie
ann.   Jun-Aug   vine-like   N
Value for honey: (for *Vicia* L.) HP3[16]; has the reputation of supplying more N, in larger quantities and for a longer period than any other plant known[54].
Notes: flowers yellowish white or purple tinged.

Vicia sativa L.
spring vetch, tare, vesce cultivée, vesce commune
ann. bienn.   Jun-Aug   less than 1 m long   N
Value for honey: HP1/2[17]; attractive for N, but not first class or very dependable[35]; produces a thick stipular (extra-floral) N which is more attractive to honeybees than floral N[17]; stipular N secretion begins approximately 10 days before floral N secretion[72].
Honey: may be dark amber, and stronger than *Trifolium* (true clover) H[17].
Notes: one of the best legumes for low fertility soil[54]; flowers purplish; occasionally spreads from cultivation to roadsides and waste lands[23].

Vicia villosa Roth
hairy vetch, winter vetch, vesce velue
ann. bienn.   very hardy   Jun   20-30 cm   N
Value for honey: HP2[17]; H yield up to 45 kg/colony/season[17]; considered more erratic in N production and attraction than *V. sativa* (common vetch)[54]; no extra-floral N production[54]; surplus H was reported from B.C., where this was grown as a cover crop in orchards[72]; was recommended in past for sowing in waste land in B.C. in order to increase bee forage[73].
Honey: light; classed as water-white to extra-light amber; rapid granulation; heavy body; flavour mild to very mild[17].
Notes: cultivated and spread throughout much of Canada, along roadsides and in waste land[23].

Vigna radiata (L.) R. Wilcz. var. *radiata*[74]
(syn. *Phaseolus aureus* Roxb.; *P. radiatus* L.)[74]
mung bean, golden gram, Oregon pea, haricot mungo
ann.   grows anywhere in the corn belt   30-90 cm   N(P?)
Value for honey: where grown as a crop, bees observed to work it freely for the entire day over a period of 3 weeks[72]; in the American Bee Journal Test Garden, it did not attract honeybees[72].
Honey: see *Phaseolus* L.
Notes: grown for beans, green manure and as a cover crop; important crop in India and probably derived from a wild sp. growing in that area[74].

Zea mays L.
sweet corn, maize, Indian corn, maïs
ann.   1-8 m   DP
Value for honey: honeydew H reported and bees may also collect sap from split or injured stems[72]; large yields of 70-90 kg noted (acreage not indicated)[72].

Value for pollen: claimed that bees also collect P[72].
Honey: yellow in colour, and may be very dark[72]; granulation said to be coarse, and flavour peculiar to very acceptable[72].
Notes: a widely cultivated crop, grown chiefly for silage and seeds; a very important grain crop in Ont.; not persistent outside cultivation.

Zinnia elegans Jacq.
common zinnia, youth and old age, zinnia élégant
ann.   Jul-Sep   90 cm   NP
Value for honey: single cvs. only are fairly attractive to bees[35].
Notes: many races and cvs. available; popular as cut flowers; grow in full sun.

5.2 Perennials

Agastache foeniculum (Pursh) Kuntze[51]
(syn. *Stachys foeniculum* Pursh[51]; *Hyssopus anisatus* Nutt.; *Lophanthus anisatus* (Nutt.) Benth.[51]; *Agastache anethiodora* Nutt. ex Britt.[49])
anise hyssop, blue giant hyssop, fennel giant hyssop, fragrant giant hyssop, wonder honey plant
per.   hardy   Jun-Sep   60-90 cm   N(P?)
Value for honey: N sugar concentration 44%; yields all day and in cool windy weather[72]; may be preferred by bees to *Melilotus* (sweet clover), and has been source of surplus near Edmonton, Alta., and Winnipeg, Man.[72]; was reported to be the most attractive plant to honeybees[94]; one beekeeper claims that honeybees will visit no other source while this is in bloom[72]; bulk seed was available in the 1870's and this was experimentally planted as bee forage on a field scale[72].
Honey: light in colour; heavy body; flavour delicate and minty[72].
Notes: foliage has scent of anise[23]; grown as an ornamental and due to large size may be used to advantage for bold effects; dry thickets, plains and barrens, Ont. to Mackenz., S. beyond our range; E. to fields in Que., N.E.[23]; *A. nepetoides* (L.) O. Kuntze (giant hyssop, yellow giant hyssop) occurs naturally in e. Canada, but opinions differ regarding its importance[72].

Ajuga genevensis L.
Geneva bugle, bugle de Genève
per.   -50° to -37° C   May-Jun   15-25 cm   N(P?)
Value for honey: HP5[16].
Notes: flowers well in sun or part shade; grows in clumps; use in a border or rockery; reproduces by runners; locally in fields, waste land, N.E. to Ont., S. to beyond Can. limits[23].

Ajuga reptans L.
(syn. *A. repens* N. Tayl.)
bugleweed, carpet bugle, bugle rampante
per.   -50° to-29° C   early May-mid Jun   10-30 cm   N(P?)
Value for honey: HP5[16].
Notes: vigorous and can be invasive; useful on banks for ground cover; sun or shade; locally along roadsides and in fields, Nfld. to Wisc., S. beyond Can. limits[23].

Allium cepa L. and *Allium* L. (see individual spp. for more detail)
onion, chives, leek, garlic, ail, oignon, ciboule
per.   bulb   Jun   N(P)
Value for honey: (for *A. cepa* L., onion) all Allium spp. are generally assumed to have similar value, HP3[16], but several *Allium* spp. (not *A. cepa* L., onion) do not have nectaries[49]; one source reports that onions produce 40% less per day as sainfoin[54]; super-phosphate fertiliser may increase N sugar concentration, but not volume[54].
Bees may sometimes be reluctant to forage on onions[54]. N yield is better on moist sites. Each flower head consists of a large number of florets that bloom over a 2-3 week period so that each plant yields for a long time[54].

Original prairie flora included wild onions which were known to be an important source of H for the early beekeeping settlers[72].

Except where vegetable *Allium* are grown for seed, these crops are harvested before flowering and are of no use to the beekeeper.

Value for pollen: less attractive for P than N.

Honey: light amber in colour; good quality; oniony flavour present only in freshly extracted H and disappears during storage[16].

Notes: strong scented and pungent herbs; several ornamental *Allium* have been specifically noted to be attractive to honeybees, but many other native or naturalized spp. are probably also valuable for N; see also *A. porrum* var *porrum*[51](leek, poireau), *A. sativum* (garlic, ail), and *A. schoenoprasum* (chive, ciboulette) listed in Table 19.

Allium giganteum Regel
giant onion
per.   -23° to-21° C   Jun   120 cm   N(P)
Value for honey: yields N copiously over a 2-3 week period; see *Allium*.
Honey: see *Allium*.
Notes: needs staking; moist sites; some material grown under this name is *A. ameloprasum* L.; see *Allium*.

Althaea officinalis L.
marshmallow, white mallow, guimauve officinale
per.   -23° to -20°   Jul-Oct   1-1.8 m   NP
Value for honey: HP3/4[16]; attractive for N[35]; probably similar to *Alcea rosea* (hollyhock).
Notes: large pinkish flowers on a stalk; root yielding the original non-synthetic mucilaginous marshmallow paste[23]; naturalized along borders of saline or fresh marshes, especially in E., very local, Co. Deux-Montagnes, Que., also sw. beyond Can. range[23].

Anaphalis margaritacea (L.) Benth. & Hook.f.
common pearly everlasting, pearly everlasting, immortelle, anaphalide nacrée
per.   -37° to -29° C   Jul-Aug   90 cm   NP
Value for honey: H yield of 9 kg/colony/season reported[72]; surplus usually obtained when this sp. is near colonies sited on fireweed (*Epilobium angustifolium*) and the fireweed main N flow fails[50]; main N flow follows the fireweed flow[72].
Honey: dark in colour; may be thin; flavour is fair to bitter; acceptable for over-wintering[50,72].
Notes: widespread in N. America; extremely variable appearance.

Anchusa azurea Mill.
(syn. *A. italica* Retz.)
Italian bugloss, Italian alkanet, bugloss, alkanet, buglosse
per.   -37° to -29° C   Jun-Sep   90-150 cm   N[36]
Value for honey: may be similar to *A. officinalis* (alkanet, bugloss) which has an HP of 2/6[16], but values for *A. azurea* have not been measured.
Notes: beautiful blue flowers; easy to grow, roadsides and waste land, local N.E. to sw. beyond Can. range[23].

Anemone blanda Schott R. Kotschy
Greek anemone, anémone
per.   -21° to -15° C   May   10-15 cm   P
Notes: sun or part shade; well-drained site.

Anemone nemorosa L.
European wood anemone, anémone, anémone sylvie
per.   -37° to -29° C   Apr-May   20 cm   P[36]
Notes: white or purple single flowers

Anemone patens L.
(syn. *Pulsatilla patens* (L.) Mill.)
spreading Pasque flower, prairie crocus, anémone pulsatille
per.   -23° to -21° C   Apr-May   30 cm   P
Value for pollen: one of the first herbaceous plants to supply P on the prairies (found from Man. to Alta.)[72].
Notes: not commonly cultivated; has been used medicinally; prairies and exposed slopes, Arct., nw. USA. and Can., S. to beyond Can. limits[23].

Arabis caucasica Schlechtend.
wall rock cress, arabette du Caucase
per.   -37° to -29° C   Mar-Jul   15-25 cm   NP
Value for honey and pollen: one of the earliest flowering plants to yield N and P so freely.
Notes: useful as a ground cover plant; needs sun or part shade.

Armeria maritima (Mill.) Willd.
(syn. *Statice maritima* Mill., *S. armerica* L. in part; *Armeria vulgaris* Willd.)
common thrift, armérie maritime
per.   -37° to -29° C   mid May-mid Jun   15-30 cm   N
Notes: grows well in waste land; requires sun.

Asclepias L.
milkweed, locoweed, silkweed, herbe à ouate, asclépiade
per.   Jun-Oct   N(P?)
Value for honey: HP1/2[16]; many of these are considered valuable N sources[28,72]; important where persistent in fields, ditches, roadsides etc.; capable of contributing a significant amount of N to the honey flow when most other summer blooming plants are at an end[28].

Sometimes N collecting bees fatally trapped in flowers by strands attached to pairs of "pollinia" (P structures), characteristic of milkweeds. Several masses of pollinia may be inadvertently carried on a bee's body from flower to flower. Bees may be unable to remove the pollinia from their bodies and strands may become entangled in a milkweed flower. As a result, bees that are unable to escape starve to death[28].
Value for pollen: P grains are bound together in a waxy mass and probably not actively collected[50]; P may be in a form that makes it unavailable for storing by bees[50] and P structures may fatally entangle bees[50] (see Value for honey above).
Honey: light in colour, water-white from some spp.[16]; excellent and mild flavour; cappings pearly-white and ideal for comb H production[28]; hot dry weather makes pure H very thick and difficult to extract unless warmed first in combs[28].

Notes: about 50 spp. native to N. America, some very limited in distribution; range of some spp. reduced by intensive cultivation, and now commonly found in fields which are least disturbed by arable farming or on railway embankments and unsprayed roadsides[23].

A characteristic of this group of plants, is the milky-white sap that "bleeds" from a cut stem or leaf. Also known for the distinctive mature seed-pods, packed with seeds bearing long silky hairs. Frequently insects feeding on milkweeds are poisonous to mammals.

Ornamental spp. are grown in the border or rock garden (e.g. *A. tuberosa* L., butterfly milkweed, asclépiade tubéreuse). Common native American spp. include, *A. syriaca* L. (common milkweed, herbe à coton) found in thickets, dry fields and roadsides in w. N.B. to Sask. and S.[23]; *A. incarnata* L. (swamp milkweed, asclépiade incarnate) found in wet thickets, swamps and shores in Que., N.S., Man., N.E., L.I. and beyond Can. limits[23]; *A. verticillata* L. (whorled milkweed) found in dry woods and open sterile soil in Mass., N.Y., s. Ont., Mich., Wisc., s. Man., and s. Sask.; *A. speciosa* Torr. (showy milkweed) found in prairies and clearings in Minn. to B.C.[23].

Aster L.
aster, Michaelmas daisy, starwort, frost flower
per.   Aug-frost   N(P)
Value for honey: HP2 (for most *Aster*)[16]; H yields of 13.5-45 kg/colony/season reported[50]. Asters are one of Canada's major naturally occurring source of N and P. Most apiaries in Canada are within easy reach of at least one sp. of aster. However, few beekeepers differentiate between spp. when reporting, so it is difficult to compare them accurately.

Asters are considered valuable for their long late flow and ability to yield well, even in very cool weather. The apparent variation in value for H between spp. may be due to differences in growing conditions. The aster is found in all provinces, but is commercially more important to beekeepers in the E. (i.e. Ont., Que., and the Maritimes)[50,72].

Value for pollen: much less important for P than N.

Honey: surplus H is often mixed with that of *Solidago* (goldenrod), but since the latter is more sensitive to frost than asters, the production of pure aster H is sometimes possible[72].

H is light to medium amber in colour; with a characteristic aroma; unripe H may have a rank or sickening odour which largely disappears during storage; usually of fine quality, but can have characteristics making it undesirable for over-wintering (i.e.: contains "gums" indigestible to bees; early granulation; and prone to fermentation due to late collection and incomplete ripening). However, in some areas it is depended upon for winter stores[72].

Many of the garden cvs., hybrids and native spp. are probably worked well for N and P, but recorded evidence was found only for those spp. listed here or described below: *A. ericoides* L. (Me., s. B.C.); *A. sagittifolius* Wedem. ex Willd. (N.B. to Ont.); *A. sedifolius* L.; *A. tradescantii* L. (s. Nfld., L.St.John, Que., N.S., Bruce Pen., Ont.); *A. umbellatus* Mill. (Maritimes to Man.); *A. vimineus* Lam. (e. Canada)[23,72].

Notes: Other notable native aster spp. of undetermined value for H are: *A. macrophyllus* (Que., Ont., N.S.); *A. azureus* Lindl. (Ont.); *A. ciliolatus* Lindl. (Anticosti I. to Hudson Bay reg., nw. to n. B.C., S. to N.S.); *A. undulatus* L. (N.S., s. Ont.); *A. tardifolius* L. (n. N.E., s. Que., e. Ont.); *A. sericeus* Venten. (s. Man.); *A. radula* Ait. (Nfld., Que., N.S.); *A. falcatus* Lindl. (Alta.) *A. simplex* Willd. (w. Nfld., Sask.); *A. adscendens*

Lindl. (w. Nfld., Gaspé Pen., Que., Manitoulin Distr., Ont., Sask. to Alta.); *A. foliaceus* Lindl. (Que., N.S., N.B.); *A. johannensis* Fern. (s. Lab., Sask., S. to Nfld., P.E.I., N.B., s. Ont.)[23].

Aster amellus L.
Italian aster, aster amelle
per.   -37° to -29° C   Aug-Sep   45-60 cm   N(P)
Value for honey: several cvs. available that may vary in attractiveness, but the sp. is known to be valuable[36]; see *Aster*.
Honey: see *Aster*.
Notes: purple flowers, 30 mm across are large and showy; more difficult to grow than most asters; introduced to N. America.

Aster cordifolius L.
blue wood aster
per.   Sep-Oct   90-120 cm   N(P)
Value for honey: a reported source of surplus in the Gâtinais Valley in Que., where it is used successfully for over-wintering[72]; see *Aster*.
Honey: unlike H from most asters, this does not granulate readily; see *Aster*.
Notes: many cvs. available that may vary in value; open woods, thickets, and clearings, usually common, Que. to Wisc., N.S., N.E. and beyond Can. limits[23].

Aster dumosus L.
bushy aster
per.   -50° to -29° C   Aug-Oct   90 cm   N(P)[36]
Value for honey: produces H freely[72]; see *Aster*.
Honey: see *Aster*.
Notes: flowers prolifically; found in parts of the e. USA.; dry to wet siliceous and argillaceous open ground, thickets, shores, s. Me., and beyond Can. limits[23].

Aster laevis L.
smooth aster
per.   -29° to -23° C   Sep-Oct   NP
Value for honey: N sugar concentration 41%[94]; may be more attractive for P than N[94].
Honey: see *Aster*.
Notes: grows well in dry soil[95]; a good plant for prolonging the blooming season in the garden; white, pink or purple flowers in terminal clusters; Yukon to ne. Ore. and E. to Me.[23] (and N.S.?).

Aster lateriflorus (L.) Britt.
calico aster
per.   -37° to -29° C   Aug-Oct   120-150 cm   N(P)
Value for honey: produces N freely[72]; see *Aster*.
Honey: see *Aster*.
Notes: grows in moist or dry locations; flowers are 30 mm across and borne in one-sided clusters on the branches; native to se. Canada; dry to moist fields, clearings, thickets, shores etc., Algoma Distr., Ont., S. to N.S., N.E. and beyond Can. limits[23].

Aster novae-angliae L.
New England aster, aster de la Nouvelle Angleterre
per.   -50° to -29° C   Aug-Oct   90-150 cm   N(P)
Value for honey: N sugar concentration of 25%[94] (fairly dilute); despite a thin N, it is very attractive to honeybees and there is often more than one foraging bee per flower at any one time; greedily sought-after especially when air temperatures are cool[94]; when collecting N, honeybees and native bees may show a peculiar behaviour, as if intoxicated[72]; see *Aster*.
Honey: see *Aster*.
Notes: a common wild plant from Que. to Alta., and is also commonly cultivated; requires a moist site; damp thickets, shores, meadows etc., sw. Que. to s. Alta., S. to centr. Me., s. N.E. and beyond Can. limits; frequently cultivated and also escaped from cultivation[23].

Aster novi-belgii L.
New York aster, aster de la Nouvelle Belgique
per.   -50° to -29° C   Sep-Oct   90-150 cm   N(P)[36]
Value for honey: see *Aster*.
Honey: see *Aster*.
Notes: many cvs. available which make excellent ornamentals; moist site; native to many coastal sites from Nfld. and S.; damp thickets, meadows, shores etc., mostly within 160 km of the sea from Nfld. to s. Que. and S. to N.S., s. N.E. and beyond Can. limits[23].

Aster puniceus L.
swamp aster
per.   -37° to -29° C   Sep   120-240 cm   N(P)
Value for honey and pollen: an important N and P source in the Maritimes, Que., and Ont.[72]; see *Aster*.
Honey: see *Aster*.
Notes: due to its extreme height, this aster may be best in a "wild" garden; prefers a moist to swampy site; a var. of this found in swamps and damp thickets in Nfld., s. Man., N.S., N.E. Thunder Bay Distr. Ont., Gaspeé Co. Que., Lab. to Mackenz. Distr. etc.[23].

Astralagus cicer L.[51]
cicer milkvetch, mountain chick pea, astragale pois chiche
per.   N(P?)
Value for honey: attractive to honeybees, but specific value is unknown[54]; *A. sinicus* L.[51], Chinese milkvetch, yields 30 kg/colony/season[72].
Notes: a legume that may be useful in irrigated and dryland areas as a soil improver or forage crop; not commonly cultivated in N. America.

Aubrieta deltoidea (L.) DC.
purple rock cress
per.   -29° to -23° C   late Apr-Jun   8-15 cm   NP
Value for honey: yields N freely and is one of the earliest per. to bloom[35]; shear off stalks after flowering to encourage second fall bloom.
Notes: does well in parts of B.C.; less easily grown in e. Canada; sun or part shade.

Aurina saxatilis (L.) Desv.
(syn. *Alyssum saxatile* L.; *A. orientale* Ard.; *A. arduini* Fritcsh)
basket of gold, rock madwort, goldentuft, corbeille d'or
per.   -37° to -29° C   May   30 cm   (N?)P
Value for honey: opinions differ about attractiveness to honeybees for N[35]; infrequently visited in the author's garden (U.K.).
Value for pollen: attractive for P[35].
Notes: used in rock gardens for the show of bright yellow flowers; attractive bluish green leaves.

Baptisia australis (L.) R. Br.
blue false indigo, plains false indigo, wild blue indigo
per.   -59° to -29° C   May-Jun   90-120 cm   N
Value for honey: has proven very attractive in the Pellett Test Garden[72].
Notes: pea-like blue flowers and rich bluish green foliage; clumps of this can make an attractive summer-time hedge[95]; found naturally in rich woods and alluvial thickets[23].

Baptisia tinctoria (L.) Venten.
wild indigo, yellow indigo, horsefly weed, rattleweed, baptiste des teinturiers
per.   -23° to -21° C   Jul   100-120 cm   N
Value for honey: said to yield well under all conditions; N flow may be intense[72].
Honey: light amber in colour; flavour is pleasant and characteristic.
Notes: yellow pea-like flowers; requires little care, and can survive droughty conditions[95]; a dye plant.

Caltha palustris L.
marsh marigold, cowslip, may blob, souci d'eau, populage
per.   -37° to -29° C   Apr   30-90 cm   NP
Value for pollen: noted to be an important source of P in the Maritime provinces; also reported to be toxic to honeybees on occasion, probably due to alkaloids in the P[7]; see Section 10.
Notes: grows in areas which are wet in the spring but become dry later in the summer; dies to the ground after flowering; flowers resemble buttercups; herbage sometimes used as a pot-herb; found in swamps, wet meadows and wet woods, Lab. to Alaska, s. to Nfld., N.S. and S. beyond Can. limits[23].

Camassia Lindl.
camass, camas, quamash
per. (bulb)   -23° to -21° C   May-Jun   NP[36]
Value for honey and pollen: attractive to honeybees[35]; *C. cusickii* S. Wats. (Cusick camas); *C. quamash* (Pursh) Greene (common camas); and *C. scilloides* (Raf.) V.L. Cory (wild hyacinth, indigo squill) are all noted to be freely visited by honeybees[35,72].
Notes: useful for naturalizing in heavy moist soil; found in low fields, open woods and meadow, s. Ont., se. Mich., s. Wisc. and beyond Can. range[23].

Campanula carpatica Jacq.
Carpathian bellflower, tussock bellflower, campanule des Carpathes
per.   -37° to -29° C   Jun-Sep   15-45 cm   NP[36]
Value for honey: vigorously worked for N on occasion[35].
Notes: spreads, but is not weedy; sun or part shade.

Celastrus scandens L.
American bittersweet, false bittersweet, climbing bittersweet, bourreau des arbres, célastre grimpant
per.   zone 3b   Jul-Sep   vine-like, to 60 cm   N
Value for honey: bees work freely on this[72].
Notes: yellow clusters open to disclose red berries; often used for decoration[95]; thickets, river banks and woods, s. Que. to s. Man., S. to s. N.E. and beyond Can. limits[23].

Centaurea dealbata Willd.
Persian centaurea, Persian knapweed
per.   -37° to -29° C   Jun-Sep   60 cm   N
Value for honey: said to be particularly attractive to honeybees[72]; see *Centaurea*.
Honey: see *Centaurea*.
Notes: solitary flower heads, white and rosey-white; a compact bushy plant.

Chionodoxa luciliae Boiss.
glory of the snow, gloire de neige
per. (bulb)   Apr-May   7-15 cm   N(P)[36]
Value for pollen: a useful early source of P[35].
Notes: needs full sun; will naturalize; among the first spring bulbs to bloom; small blue and white bell-shaped flowers in a cluster; foliage is grass-like.

Cichorium intybus L.
common chicory, blue daisy, blue sailors, wild succory, chicorée, chicotin, barbe du capuchin, chevaux de paysan
per.   -37° to -29° C   90-180 cm   N
Value for honey: HP3[16]; plants may close up in the afternoons and so yields N in the mornings only[35]; surplus H has been obtained where this is grown on a field scale for roots or seeds (e.g. England and Mich.)[35,72]; one of the most attractive plants to bees[72].
Honey: is yellowish-green in colour; flavour is pleasant and has a coffee-chicory taste[72].
Notes: cultivated as a crop for roots or green fodder in some countries[35], but is a noxious weed is some provinces of Canada[3].

Cirsium Mill.
thistle, plume thistle, chardon
per. (bienn.)   Jul-frost   30-150 cm   N
Value for honey: HP3[16]; most of these are excellent sources of N; however, some thistles have very long flower tubes so that the N is out of reach of the honeybee, or rarely available[35]; *C. arvense* (L.) Scop., Canada thistle, is one of the best thistles for N, but is a serious introduced pest or noxious weed in several provinces[35,72]; *C. crispus* (welted thistle) is also a weed but is eagerly visited by honeybees[35].
Value for pollen: may be eagerly collected (especially from *C. arvense*).
Honey: light in colour; good flavour; no pronounced aroma[16]; has been compared to basswood (*Tilia*) H in characteristics and quality[72]; some state that it is the finest available source of H[72].
Notes: most of these are serious weeds due to their invasive nature, and are usually unwelcome intruders in the garden. Flowers are not unattractive and are made up of hundreds of tiny tubular florets in tight globular clusters.

Claytonia virginica L.
Virginia spring beauty, spring beauty, claytonie
per. (bulb)   -21° to -15° C   Apr-May   10-15 cm   NP
Value for honey and pollen: valuable only because they are so early flowering when little else is in bloom[72].
Notes: pink flowers, useful ground cover in part shade; moist soil; native to e. Canada[72]; rich woods, thickets, and clearings, sw. Que., and w. and s. N.E. to s. Ont. and beyond Can. limits[23].

Clematis L.
clematis, virgin's bower, leather flower, vase vine, clématite
per.   vine   (N)P
Value for honey and pollen: only some offer N, but all are sources of P except in the case of some of the garden hybrids (e.g. × Jackmanii T. Moore which is useless for bees[35]); *C. virginiana* L. (white clematis, woodbine, virgin's bower, devil's-darning-needle, herbe aux gueux) is not often seen in gardens, but produces abundant N[72] and is common in N.S., Que., Man. and locally in other provinces.
Notes: The clematis flower has no true petals, but the brilliantly coloured sepals serve instead. This large group of useful vines are not as widely appreciated in N. America as they are in Europe where many of the smaller flowered spp. (unknown in most N. American gardens) are popular. Many clematis spp. are probably popular with honeybees, but there are few records indicating the value of particular spp. to beekeepers.

Clematis armandii Franch.
Armand clematis, clématite du Père Armand
per.   -15° to -12° C   May-June   vine to 12 m   NP
Value for honey: observed to "hum" with bees.
Notes: white fragrant flowers in showy panicles; blooms on previous year's wood[95]; thrives in coastal B.C., but is too tender in most other parts of Canada.

Clematis ligusticifolia Nutt.
western virgin's bower, western white clematis
per.   -23° to -21° C   Aug-Sep   vine to 6 m   NP
Value for honey: a 3.6 tonne surplus reported from wild clematis (likely *C. ligusticifolia*) in B.C. in the 1930's[72]; N sugar concentration 48%, but bees observed to obtain very little N from each blossom[94].
Notes: white flowers in plumose heads; climbing over bushes in Man. to B.C., S. beyond Can. range[23].

Clematis montana Buch.-Ham. ex DC.
anemone clematis, clématite des montagnes
per.   -21° to -15° C   Jul   2.4 m   NP
Value for honey and pollen: attractive.
Notes: grows best on s. or sw. walls; white flowers.

Clematis paniculata J.F. Gmel.
sweet autumn clematis, clèmatite à paniculés
per.   -23° to -21° C   late Aug   vine to 30 m   NP
Value for honey and pollen: attractive[91].
Notes: white flowering exotic clematis with an excellent fruit display.

Clematis vitalba L.
traveller's joy, clématite des bois
per.   -20° to -23° C   vine   late Aug   NP
Value for honey: observed to be "buzzing" with bees[35].
Notes: flowers in white to greenish-white panicles, fragrant.

Colchicum autumnale L.
meadow saffron, autumn crocus, fall crocus, colchique, dame nue, veilleuse
per.   -29° to -23° C   Oct   10 cm   P
Value for pollen: very attractive, especially if little else is available[35].
Notes: naturalizes in rough grassy areas; dried corms and seed source of medicinal colchicum and colchicine; spreads locally to meadows and fields[23].

Coronilla varia L.
crown vetch, coronille bigarée
per.   -37° to -29° C   Jun-Sep   (N)P
Value for pollen: not a good source of N for honeybees[54]; bees have been known to starve on large acreages of this (e.g. Penn.)[54]; more attractive for P[54]; did not appear to secrete N (Guelph, Ont.)[1].
Notes: long-lived drought tolerant legume that grows in soils of very low fertility; especially valuable for holding banks; stand improves with age and chokes out weeds[54]; roadsides, waste land and old house sites, originally cultivated, now abundantly but locally established[23].

Crocus L.
crocus
per. (bulb)   Apr or Sep   5-12 cm   (N)P
Value for honey: N is sometimes available if it accumulates and rises sufficiently in the long flower tubes for the honeybee to be able to reach it; usually more attractive for P[72].
Value for pollen: abundant supply of bright orange P which is high in protein; beekeepers usually contend that the orange (or dark yellow) crocus is more attractive to honeybees than crocuses of other colours[35]; chilled crocus P may be toxic to bees[7] (see Section 10).
Notes: naturalizes in lawns or under trees; many spp. available, either spring or autumn flowering.

Dahlia Cav.
dahlia
per. (tuber)   tender   Jul-frost   45-75 cm   NP
Value for honey and pollen: several bees may work on blossom at once[72]; dwarf bedding singles last the longest and therefore are the best choice[35].
Notes: tender tuberous plants which have been garden favourites; single cvs. only visited by bees.

Dictamnus albus L.
burning bush, gasplant, dittany, fraxinella, fraxinelle
per.   -50° to -21° C   Jun   60-90 cm   NP[35]
Value for honey and pollen: useful to bridge the June gap.
Notes: a favourite in the flower border of bygone days; purple flowers; full sun; aromatic leaves; foliage emits an ethereal inflammable oil.

Doronicum plantagineum L.
leopard's bane, plantain leopard's bane, doronic plantain
per.   -37° to -29° C   May-Jun   120 cm   NP[36]
Notes: yellow sunflower-like flowers; sturdy and of easy culture.

Echinops exaltatus Schrad.
Russian globe thistle, échinope de Russie
per.   -37° to -29° C   90-360 cm   N
Value for honey: considered equally attractive as *E. sphaerocephalus* (globe thistle)[35].
Notes: the best of any of the spp. in this genus for garden planting[95]; compound blue flower-heads.

Echinops ritro L.
small globe thistle, boulette azurite, petite boulette
per.   -27° to -29° C   30-360 cm   N
Value for honey: considered equally attractive as *E. shaerocephalus* (globe thistle), but is a more desirable ornamental.
Notes: smaller than most globe thistles; bright blue compound flowers.

Echinops sphaerocephalus L.
globe thistle, Chapman honey plant, great globe thistle, boulette azurée, boulette commun
per.   -37° to -29° C   mid Jul-end Aug   90-240 cm   N
Value for honey: HP5[16]; noted to be an excellent source in some areas[72]; claimed to be very valuable for N when grown experimentally in Man. in the late 1940's[72]; also was grown on trail in N.Y. State by H. Chapman to prove its attractiveness (some observers contended that bees only idled on the blossoms and collected little N)[72].
Notes: grown as a forage crop in some countries; can be grown as a summer-time apiary hedge if it is supported by two wires (dead stems can be left in position as a winter windbreak)[35]; sun and well-drained soil; may be listed in nursery catalogues, but other spp. may be offered under this name; naturalized in local areas of N. America; frequently cultivated and occasionally a volunteer plant on waste heaps, in fields and other waste places, s. Que. and beyond Can. limits[23].

Epilobium angustifolium L.[51]
(syn. *Chamaenerion angustifolium* (L.) Scop.; *Epilobium spicatum* Lam.)[51]
fireweed, rosebay willowherb, Indian pink, wickup, épilobe en épi, osier de St. Antoine, Antionette
per.   -37° to-29° C   Jul-Aug   2.5 m   N(P)
Value for honey: HP5/6[16]; exceptional HP of 1000 kg/ha recorded[17]; H yield of 25-57 kg/colony/day[17]; a 10 year average yield is likely to be a disappointing figure, since the N flow is not dependable from year to year[72]; N flow is heaviest during the hot weather, when the air is clear, humid and still[17]; N secretion is optimum at

temperatures between 23° to 25° C, RH 60-70[17]; maximum secretion occurs between 1800 and 0600h, but sugar content is highest from 1000 to 1400 h[17]; heaviest yields are likely in the first few years following a forest fire, and before the plants are crowded out by other more competitive pioneer spp.[73]; grows luxuriantly in the Fraser Valley, B.C. but not abundant enough to be valuable to beekeepers[a] ; yields well and is a major source of surplus, in West Kootenay area of B.C. and some parts of Vancouver I.[73]; no other major H plant grows as far north in Canada[72].

Honey: white in colour[17]; granulates to a fine to medium grain[17], but may remain liquid for several years; flavour described as mild, rich and almost none, but always very sweet[17].

Notes: variable boreal spp.[23]; handsome plant with rosey flowers on a tall spike; spreads by underground rhizomes; growth and spread can be encouraged by applying nitrogen fertilizers (at certain application rates this may also increase N secretion); full sun; does not compete well with other invasive plants; found in recent clearings, burned woodlands, damp ravines, subarct. America. S. to Nfld., N.S. and across Canada and beyond Can. limits[23]; several var. of the spp. are recognized (e.g. var. *macrophyllum* (Haissk.) Fern.[23], local w. Nfld., M.I.; var. *intermedium* (Wormsk.) Fern.[23], Greenl. and Lab. to Nfld., E. to Que., L. Mistassini and James Bay[23]; var. *platyphyllum* (Daniels) Fern.[23], Shicshock Mts., Que., B.C. and S.[23]); see Table 18.

Eranthis hyemalis (L.) Salisb.
winter aconite, hellebore d'hiver, eranthus d'hiver
per. (tuber)   -29° to -23° C   Feb-Mar   10 cm   (N)P
Value for pollen: an excellent source of early P if planted near colonies.
Notes: blooms last 4-6 weeks, beginning before crocuses[35]; grows well under deciduous trees and shrubs, but not in grass; once established needs little attention; yellow buttercup-like flowers; naturalized in N. America.

Erigeron speciosus (Lindl.) DC.
fleabane, Oregon fleabane, vergerette
per.   -58° to -37° C   60-90 cm   N
Value for honey: freely visited by bees[35].
Notes: an outstanding flowering per. for the border; violet purple flowers in aster-like clusters; native to s. B.C. and Alta. to s. Ore. and S. beyond Can. range[23].

Eryngium maritimum L.
sea holly, panicaut de mer, panicaut marin
per.   -23° to -21° C   Jul   30 cm   N
Value for honey: very attractive to honeybees when massed[35].
Notes: flowers are blue and thistle-like; will tolerate poor soil; full sun preferred; naturalized along the Atlantic coast; *E. giganteum* Bieb. is not commonly cultivated but is noted to be very attractive to honeybees[35].

Eupatorium perfoliatum L.
boneset, thoroughwort, eupatoire
per.   -37° to -29° C   Jul-Oct   150 cm   NP
Value for honey: said to be one of the best *Eupatorium* for H in Canada[72]; an important late source of N.

---

[a] Mitchell, S. (1987) *Personal communication.*

Honey: usually mixed with other autumn sources[72].

Notes: white flowers in loose terminal clusters; occurs wild across Canada and is sometimes grown in gardens; native to e. N. America[49]; found in low woods or thickets, swales, wet shores and prairies; usually common in Que to se. Man., S. to N.S. and beyond Can. limits[23].

Foeniculum vulgare Mill.
(syn. *F. officinale* All.)
fennel, common fennel, fenouil
per., grown as an ann. in Canada   Jul-Sep   150 cm   N

Value for honey: intensely visited by bees[54]; suspected to yield a more concentrated N in dry climates.

Notes: commercially grown for seeds used as a condiment; commonly grown in herb gardens; feathery foliage makes an attractive addition to the flower border; found naturally in dry fields and along roadsides[23].

Galanthus nivalis L.
common snowdrop, perce neige, clochette d'hiver
per.   -37° to -29° C   Mar-Apr   NP

Value for honey: bees visit these eagerly[35].

Notes: white flowers; among the earliest bulbs to bloom; they usually increase themselves (by seed) if left alone; partial shade is best; blooms last about 3-4 weeks[95]; spreads very slightly from cultivation[23].

Geranium ibericum Cav.
meadow cranesbill, meadow geranium, géranium
per.   -23° to -21° C   May-Jul   30-80 cm   NP

Value for honey: HP3[16]; N sugar concentration 57-71%[17]; a very good source[35]; *G. phaeum* L. (dusky cranesbill, géranium livide) and *G. sanguineum* L. (blood red geranium, sanguinaire, herbe à becquet) are also noted to be attractive to honeybees[35], but are not described here since they are not commonly cultivated in N. America.

Notes: purplish coloured flowers and deeply incised handsome foliage; not to be confused with the common garden geranium that is often grown as a house plant (i.e. *Pelargonium*); introduced sp.; *G. pratense* (meadow geranium, géranium des prés) is also an exotic, and has spread more widely than meadow cranesbill but its value to bees is unrecorded.

Geum L.
avens, benôitre
per.   -37° to -15° C   May-Jun   25-90 cm   P[36]

Notes: brilliant flowers; some may bloom to frost; well-drained soil; many native spp.

Gladiolus L.
gladiolus, glads, corn flag, glaieul
per. (corm)   very tender   60-120 cm   N(P)

Value for honey: N is very dilute, but abundant and bees gather it readily[72]; where commercially grown, it can be an important source[72].

Value for pollen: P is collected; P is purplish[72].

Notes: colourful flowers borne on a long stalk, blooming successively from base to tip; corms must be lifted to prevent exposure to frosts.

Gypsophila paniculata L.

baby's breath, gysophile paniculée

per.　　-50° to -29° C　　Jul-Aug　　90 cm　　N[36]

Notes: white flowers borne on clusters; up to 1000 flowers per multi-branched panicle; shearing back after blooming may encourage a second flowering in Sep-Oct; rock garden or ground-cover plant; cultivated and locally spread to sandy or rocky roadsides and waste places from N.E. to Man. and S. beyond Can. limits[23].

Gypsophila repens L.

creeping gysophila, gysophile rampante

per.　　-37° to -29° C　　Jun-Jul　　15 cm　　N[36]

Notes: profuse white bloom; makes an excellent ground cover; full sun and limestone soil[95].

Hedera helix L.

English ivy, lierre commun, lierre grimpant, lierre d'Angleterre

per.　　zone 6　　Oct-Nov(?)　　clinging vine to 30 m　　NP

Value for honey: HP5[16]; last important N and P source where this is naturalized[35]; N is extremely concentrated so H is very unlikely to ferment in over-wintering stores[35]; if not all the N is gathered from flowers, a sugary crust may remain on the faded blooms[35].

Honey: is greenish in colour; fine grain; pleasant aromatic odour[72].

Notes: flowers only on the older (mature) shoots, and not on the commonly grown juvenile vine.

Hedysarum coronarium L.

French honeysuckle, Spanish esparcet, sweet vetch, sulla, sainfoin d'Espagne

per.　　-37° to -29° C　　Jun-Jul　　1-2 m　　N(P?)

Value for honey: freely visited for N[25].

Honey: is very light in colour; classed as water-white to light amber; finely grained and when granulated becomes white in colour, but not hard; flavour is mild, delicate, but may be slightly acerbic (like raw green beans) in drought areas; aroma very slight[17].

Notes: ornamentally cultivated as a ground cover or as a crop for green manure or fodder; a soil improver in waste areas; full sun; drought resistant; fragrant flowers.

Helenium autumnale L.

common sneezeweed, hélénie commun

per.　　-37° to -29° C　　Aug-Sep　　180 cm　　NP

Value for honey: H yield 14 kg/colony/season[72]; a useful source for winter stores, but said to be bitter to unpalatable as table H; H should not be mixed with the main crop[72]; see Section 8.

Honey: light yellow in colour; heavy body; rapid granulation; markedly bitter flavour[72]; small amount of sneezeweed H could spoil a main crop[72]; apparently less bitter in some countries (e.g. England)[35]; see Section 8.

Notes: best in a "wild" garden or at the back of the border; generally considered a weed; found growing wild from Que. to B.C.[49]; rich thickets, meadows and shores[23].

Helianthus salicifolius A. Dietr.
(syn. *Helianthus orgyalis* DC.)
willow leaved sunflower
per.   -37° to -29° C   Sep-Oct   1.8-3.0 m   N(P)
Value for honey and pollen: a minor source of late N and P; see *Helianthus*.
Notes: well adapted to dry sites on limestone; but may not yield well unless kept moist;
probably found in parts of s. Alta., s. Sask. and s. Man.; see *Helianthus*.

Helleborus niger L.
Christmas rose, rose de Noel, rose d'hiver
per.   -37° to -29° C   Oct-May   15-30 cm   NP
Value for honey and pollen: valuable for N and P in early spring and very late fall
when little else may be blooming[35,72]; *H. lividus* ssp. *corsicus* (Corsican rose) and *H.
orientalis* (Lenten rose) are of equal value to bees and bloom a little later, in Apr.[35].
Notes: evergreen; flowers white and borne singly.

Heuchera sanguinea Engelm.
coral bells, alum root
per.   -37° to -29° C   Jun-Aug   30-60 cm   N[36]
Notes: long blooming; branched spikes of small bell-like, red to pink flowers.

Hyacinthus orientalis L.
hyacinth, common hyacinth, Dutch hyacinth, jacinth
per. (bulb)   -21° to -23° C   Apr-May   36 cm   NP[36]
Notes: popular garden bulbs; flowers have a delightful fragrance; select singles only for
bees.

Hydrophyllum virginianum L.
Virgina waterleaf, Shawnee salad, Indian salad, John's cabbage, hydrophylle
per.   -29° to -23° C   Jun-Jul(?)   30-60 cm   N(P?)
Value for honey: very attractive for N; flowers after the fruit bloom (*Prunus*) and is
the source of surplus H where it is common[72].
Notes: needs moist soil; purple flowers in a loose umbel; foliage used as a pot-herb;
tender shoots eaten by N. American Indians; found growing wild from Que. to Man.;
rich woods and damp clearings, S. to w.N.E. and beyond Can. range[23].

Hyssopus officinalis L.
hyssop, garden hyssop, hysope officinale, herbe sacrée
per.   -50° to -29° C   Jun-Sep   20-60 cm   N(P?)
Value for honey: HP4/5[16]; although the corolla is 10 mm long, it widens at its mouth
into a funnel shape which allows the honeybee access to all the N[35]; extremely
attractive to bees.
Notes: a fragrant introduced herb that is used for perfumery as well as in cooking[35];
best grown in mass and looks attractive as an edging plant[35]; propagated easily by
seeds or cuttings; dry pasture and roadsides, locally abundant s. and w. N.E. to Ont. and
S. beyond Can. range[23]. Sometimes incorrectly referred to as *H. vulgaris* and should
not be confused with anise hyssop, *Agastache foeniculum*.

Lathyrus sylvestris L. 'Wagneri'
Wagner pea, (cv. of narrow leaved everlasting pea or flat pea, cv. de grand gesse or cv.
de gesse sylvestre)
per.   climbing to 2m.   N(P?)
Value for honey: profuse bloom and bees work this "industriously"[35]; developed
around the turn of the century by a A.W. Wagner from the narrow leaved everlasting
pea, *L. sylvestris*, to improve the value of the sp. as bee and livestock forage[72].
Notes: deep rooted and drought resistant; introduced plant; flowers dark red; sp. (not
cv.) found along roadsides and waste places, local Que. to Mich. and beyond Can.
limits[23].

Lathyrus tuberosus L.
ground nut pea, earth nut pea, tuberous vetchling, gesse tubereuse
per.   -23° to -21° C   Jun-Aug   N(P?)
Notes: introduced, rose coloured sweet-pea-like flowers; spreads vigorously and may be
suitable for banks or roadsides for erosion control. Tubers edible; fields, meadows and
roadsides in local areas of Vt. to s. Ont. and Wisc.[23].

Lavatera thuringiaca L.
(syn. *Malva thuringiaca* (L.) Vis)[51]
Siberian rose mallow, lavatera, lavatère de Thuringe
per.   (N)P
Value for honey: some N collected[16].
Value for pollen: abundant and attractive P[35].
Notes: bushy plant with hollyhock-like flowers.

Leonurus cardiaca L.
common motherwort, agripaume, cardiare, cardiaque, herbe aux tonneliers
per.   -37° to -29° C   Jul-Sep   60-120 cm   N
Value for honey: HP5[16]; N sugar concentration 39-44%[17]; N secretion usually
unaffected by drought[17]; covered with bees while blooming and reported to be more
attractive to honeybees than *Phacelia* (phacelia) or *Melilotus* (sweet clover)[35]; flower
tubes are 4 mm long so that all N is within reach of the honeybee[35].
Honey: light or straw coloured; good flavour[50,72].
Notes: common weed along roadsides; roots are very tenacious; sometimes cultivated in
the flower border; naturalized in waste land, Que., W. and S. beyond Can. limits[23].

Leonurus cardiaca L. ssp. *villosus* (Desf. ex Sprengel) Hyl.[17]
(syn. *Leonurus quinquelobatus* Gilib.; *Caridaca vulgaris* Moench; *Leonurus glaucescens*
Bunge)[51]
motherwort, léonure
per.   Jun-Jul   60-120 cm   N
Value for honey: HP6[17]; N sugar concentration 42-76%.
Honey: light to light orange in colour[17].

Leucojum vernum L.
spring snowflake, niveole-printanière, grelot blanc
per. (bulb)   -29° to -23° C   Apr-May   30 cm   NP[36]
Notes: will spread if left undisturbed; flowers white tipped with green.

Liatris pychnostchya Michx.
Kansas gayfeather, blazing star
per.   -37° to -29° C   Jul-Sep   90-120 cm   N[72]
Notes: purple flower heads in a dense terminal spike; striking appearance; damp prairies Wisc. and S. beyond Can. range[23].

Liatris spicata (L.) Willd.
spike gayfeather, blazing star
per.   -37° to -29° C   Sep   30 cm   N[72]
Value for honey: a useful late source.
Notes: grows well in dry soil; purple flowers in clusters about 50 cm long; meadows, borders of marshes, damp shores sw. Ont., Mich., Wisc. and cultivated locally and naturalized in s.N.E.[23].

Limonium latifolium (Sm.) O. Kuntze
statice latifolia, wide leaf sea lavender, marsh rosemary, statice à   larges feuilles
per.   -37° to -29° C   Jul-Sep   75 cm   N
Value for honey: said to be attractive for N; beekeepers report surplus H from *L. vulgare* Mill. (sea lavender, HP2[16]) and *L. carolinianum* (Walt.) Britt.[72].
Honey: is light in colour and pleasantly flavoured, but mild[72].
Notes: an outstanding per. for the border; bright mauve flowers in feathery panicles; popular for use in dried arrangements.

Linum flavum L.
golden flax, yellow flax, lin jaune
per.   -23° to -21° C   late Jun-mid Aug   30 cm   NP
Value for honey: similar to *L. usitatissimum* L. (flax)[35].
Notes: yellow flowers; not permanent in Ottawa; requires full sun.

Linum perenne L.
perennial flax, lin
per.   -29° to -23° C   Jun-Sep   30 cm   NP
Value for honey: similar to *L. usitatissimum* L. (flax)[35].
Notes: popular border plant; deep blue flowers in a much-branched panicle; prefers full sun.

Lotus corniculatus L.
bird's foot trefoil, bacon and eggs, pied de poule, cornette lotier, lotier cerniculé
per.   Jun-Aug   60 cm   NP
Value for honey: HP2[16]; H yields of 45 kg/colony/season[50]; N sugar concentration 14-41%[17]; has been observed to be more attractive than *Melilotus* (sweet clover) on occasion[72]; some doubt regarding value as a N source in Ont., but observation of test hives in Guelph, Ont. suggest that it is a major producer in Jul-Aug[1].
Value for pollen: load is light gray to light brown[17].
Honey: light greenish in colour; classed as white; granulates rapidly; heavy body; flavour like clover[17].
Notes: flowers about 2 weeks before white clover; grown as a crop for pasture or hay; may be suitable for roadsides, as a ground cover or planting in waste land for bee forage; flowers yellow tinged with red; fields roadsides and waste land, locally Nfld. to Minn., and S. beyond Can. limits[23].

Lupinus L.

lupine, lupin

per. (ann.)   P

Value for pollen: generally considered to be nectarless and of little value to the beekeeper, even for P[35]; *L. angustifolia* L. (narrow leaved or blue lupine) is grown on a field scale for grazing or soil improvement and is an attractive source of P (and N?) for honeybees[72].

Notes: a legume which thrives on any well-drained soil, even poor sandy loams; Latin name from *lupus*, wolf, because of an erroneous belief that this sp. destroys the soil. *L. perennis* L. var. *occidentalis* S. Wats.[23] is more common than other sp. in s. Ont.; *L. polyphyllus* Lindl.[23] is most common along dry roadsides and banks in P.E.I. and N.S.; *L. nootkatensis* Donn. ex Sims.[23] is common to roadsides and open banks Avalon Pen., Nfld. to N.E.[23].

Lythrum salicaria L.

purple loosestrife, spiked loosestrife, salicaire, bouquet violet

per.   -37° to -29° C   Jun-Sep   50-150 cm   NP

Value for honey: HP5[16]; H yield 23 kg/colony/season[17]; an Ont. beekeeper reported a 117 kg surplus near a 200 ha field growing mostly this plant[72]; N sugar concentration 30-72%[16].

Value for pollen: found to be dominant P in H sampled from the n. bank of the St. Lawrence R. (H was sampled on a regional basis)[22].

Honey: differing reports on quality of H: colour given as light, dark yellow and as yellowish gren; classed as extra-light amber; flavour strong, aromatic, sharp[17].

Notes: long dense spikes of purple flowers; moist sites; sun; naturalized in N. America, wet meadows and river flood-plains etc., locally abundant; often aggressive, choking out native vegetation, Nfld., Que., N.S.[23], P.E.I. and probably in other provinces.

Malva moschata L.

musk mallow, mauve musquée

per.   -37° to -29° C   Jul-Aug   75 cm   NP

Notes: naturalized in N. America and common in limestone soils of e. Canada and B.C.[95]; fields and roadsides, often old gardens; one or more forms abundant from Nfld. to n.N.E., less common W. to Ont. and beyond[23]; white, pink or blue showy flowers; seeds freely; tincture made from leaves can be applied to stings for relief (place leaves in jar, cover with methylated spirit, seal for 2-3 weeks, strain, dab leaf on stings as required. Fresh leaves may also be effective)[a].

Marrubium vulgare L.

white horehound, common horehound, marrube blanc

per.   -37° to -29° C   45 cm   N

Value for honey: HP2/5[16]; N sugar concentration 27-37%[17]; bees will often prefer this herb to other attractive sources[35]; *Ballota nigra* L. (black horehound, ballote noire) is not such a good bee plant since its corolla is rather long (17 mm) for honeybees[35].

Honey: has a greenish tinge; classed as amber to dark amber; when granulated, appears dark and dirty; flavour is strong like horehound candy[17].

Notes: small whitish flowers in axillary clusters; leaves and stems are used in cough medicines or for flavouring candy[95]; grows in poor soil; naturalized in N. America,

---

[a] The author cannot *personally* vouch for this remedy.

waste land, locally N. to N.E., N.Y., s. Ont.[23].

Medicago falcata L.[51]

(syn. *Medicago falcata* (L.) Mill.; *M. intermedia* Schult.; *M. silvestris* Fries; *M. sativa* L. var. *falcata* (L.) Döll; *M. sativa* L. ssp. *falcata* (L.) Arcang; *M. subfalcata* Sinsk.; *M. sativa* ssp. *vicosa* (Reichb.) G.R. Gunn)[51]
lucerne, Siberian yellow flowered alfalfa, sickle medic, yellow lucerne, luzerne jaune, luzerne sauvage
per.   very hardy cvs. available   May-Aug   NP
Value for honey: HP1/2[17].
Notes: was considered as a candidate for roadside planting in B.C. in the past[73]; roadsides and waste land, in local areas, Mass. to Man., S. to Mich. and beyond Can. limits[23].

Medicago sativa L.

alfalfa, lucerne, lentine, sainfoin, grand trèfle
per.   hardiness dependent on cv.   May-Sep   60-100 cm   NP
Value for honey: HP5[16]; H yield 45-163 kg/colony/season[17]; at a rate of 4 colonies/ha, 22 kg of H stored, but a rate of 12 colonies/ha gave little or no surplus[54]; N sugar concentration 15-64%[17]; alfalfa does not yield well under all conditions, and vigorous growth promoted by very wet springs (or irrigation) followed by hot dry conditions, gives very good to excellent yields[54] (for instance in the Interior of B.C.); seldom yields in humid weather, and therefore is an undependable source in the Maritimes[54]; generally conditions that favour seed production favour N secretion[17,54]; N secretion is severely reduced by pests on this crop[54].

An increase in alfalfa and a decrease in alsike and sweet clover acreage in Ont. in recent years, has resulted in lower average H production in that province[88].
Notes: most important legume grown in Canada and is cultivated in all provinces; important to grow the cvs. recommended for the particular area; low tolerance for poor drainage.

Blossoms open from base to tip in one week and each floret remains open for about a week[17,54]. More or less naturalized or persistent along roadsides, in old fields and waste areas[23]. See Table 18.

Melissa officinalis L.

lemon balm, bee balm, melisse citronelle, piment des abeilles
per.   -29° to -23° C   60 cm   NP
Value for honey and pollen: limited value for N and P[35]. Once commonly used to rub on the insides of bait hives to help attract swarms[a]. Also thought to ward off stings if rubbed between the hands before working with bees[35].
Notes: common garden ornamental; needs shearing during the summer to keep it neat; leaves can be used to spice beverages; naturalized in e. USA; roadsides, open woods and waste land, escaped from cultivation, local, N.E. and S. and W. chiefly beyond Can. range[23].

---

[a] The author noted a similar herb cultivated in East Africa for the same purpose.

Mentha L.

mint, menthe

per.   Jul-Sep   90 cm   N

Value for honey: HP5[16]; among the most attractive of cultivated herbs[35]; where these are (or were) grown commercially (e.g. formerly in Mich.), surplus H has been recorded from this source[72].

Honey: classed as amber; fine granulation; sharp aroma[16].

Notes: grown as a herb or garden ornamental; moist, fertile soil in partial shade preferred; some "wild" mints provide nesting material for the alfalfa leaf-cutting bee (*Megachile rotunda*)[54]; about 600 spp. have been named, but these are probably variants or hybrids of about 25 well-defined spp. (most native to Europe)[54].

Mentha arvensis L.

field mint, menthe commun, baume des champs

per.   -29° to -23° C   Jul-Aug   60 cm   N[35]

Notes: often grown in a kitchen garden and the leaves used to spice beverages; our only indigenous N. American sp. of mint, having a variable appearance and found in most provinces.

Mentha × *piperita* L. (: *M. aquatica* × *M. spicata*)

peppermint, menthe poivrée

per.   -37° to -29° C   Aug-Oct   30-60 cm   N

Value for honey: considerable crops of mint H are secured where this is grown commercially[72] (e.g. presently grown as a crop in Ore.); H yield of 90 kg/colony/season reported from more than 160 ha of this crop (i.e. Wash.)[50]; see *Mentha*.

Honey: see *Mentha*.

Notes: vigorous perennial, spreading by stolons; flowers are purple and borne in terminal spikes; has escaped from cultivation to brooksides, wet meadows etc.; should be cut to the ground every few years to keep it neat; in a good season, mint fields yield 20 kg mint oil/ha[72]; var. *citrala* (J.F. Ehrh.) Briq., bergamot mint or lemon mint, has herbage with characteristic lemon odour when crushed.

Mentha pulegium L.

pennyroyal, pudding grass, menthe pouliot, herbage de St. Laurent

per.   -37° to -29° C   Jul-Aug(?)   prostrate   N

Value for honey: see *Mentha*.

Honey: H is sometimes used for "starting" or "seeding" when making creamed H because of its fine grain (e.g. in N.Z. is used for making creamed H)[91]; see *Mentha*.

Notes: useful as a ground cover; small blue flowers; also a medicinal herb.

Mentha requienii Benth.

Corsican mint, menthella, crème de menthe plant

per.   -21° to -15° C   Jul-Aug   8 cm   N[35]

Value for honey: see *Mentha*.

Honey: see *Mentha*.

Notes: a very delicate, tiny leaved ground clover; full sun and moist soil.

Mentha spicata L.
(syn. *Mentha longifolia* of auth., not (L.) Huds.)
spearmint, baume, menthe crépue, menthe verte
per.   -37° to -29° C   Jul-Aug   60 cm   N
Value for honey: an important source where this is naturalized[72].
Honey: see *Mentha*.
Notes: widely cultivated as a sweet herb and for essential oils; found in wet places, near habitation throughout Canada[23].

Monarda punctata L.
horsemint, bergamot, dotted mint
per.   -21° to -15° C   Jul   90 cm   N
Value for honey: H yields of 9-45 kg/colony/season reported[50]; a dependable yielder[72]; some of the other bergamots (e.g. *M. didyma* L., bee balm or Oswega tea) have corollas too long to allow the honeybee to successfully reach the N[72].
Honey: is light amber in colour; clear; flavour is decidedly minty[72].
Notes: more common in e. Canada; sometimes cultivated; flowers are yellow and purple; prefers light to sandy soil.

Muscari botryoides (L.) Mill.
common grape hyacinth, muscari botryoïde
per. (bulb)   -50° to -29° C   Apr-May   30 cm   NP[36]
Value for honey and pollen: very attractive to honeybees[35].
Notes: will colonize in grass; easy culture; found naturally in pastures, fields, roadsides and fence-rows, spread from cultivation, N.E. and beyond Can. limit[23].

Myosotis scorpioides L.
true forget me not, myosotis, ne m'oubliez pas, souvenez vous de moi
per. (short lived)   -37° to -29° C   May-Jun   30-80 cm   NP
Notes: could be mixed with the cv. *semperflorens* which blooms from Jul-Aug; in quiet water or wet ground, Nfld. to Ont., S. to N.S., N.E. also Pacific coast[23].

Nepeta cataria L.
catnip, catmint, chataire, herbe à chats
per.   -37° to -29° C   Jul-Sep   90 cm   N
Value for honey: HP4/5[16]; one of the best garden bee plants known[35]. In spite of a low N sugar concentration (i.e. 22%), bees have been observed to visit catnip even when they have access to *Melilotus alba* (white sweet clover)[94]. Each flower secretes only a small amount of N, but when massed, they invite a large number of foraging bees[94].
Honey: disagreement about colour; granulates smoothly; flavour is piquant[16].
Notes: grown as a garden ornamental and not as a herb; relished by cats as well as bees (i.e. the former like to roll in it); spreads naturally along some roadsides; flowers white and pale purple; weeds of old homesteads, gardens and waste land[23].

Nepeta mussinii K. Spreng. ex Henckl.
per.   -37° to -29° C (?)   Jun-Aug   30-45 cm   N
Value for honey: HP5[16]; similar to *Nepeta cataria* L. (catmint).
Notes: confused in the trade (i.e. by nurseries and garden centres) with what is now termed *N.* × *faassenii* Bergmans ex Stearn (Persian ground ivy), but differs in being

much more sprawling in habit[95].

Oenothera L.
evening primrose, sundrops, oenothère
per.   -37° to -23° C   Jun-Sep   30-120 cm   P
Value for pollen: an early morning source of P (blossoms fade toward noon)[35]; in most spp. new blossoms open in the evening (moth pollinated) but a few spp. are day opening[35].

Onobrychis viciifolia Scop.
sainfoin, holy clover, esparcette
per.   not quite as hardy as alfalfa   Jun-Aug   10-80 cm   NP
Value for honey: HP[16], HP6[17]; H yield 20-54 kg/colony/season[17]; N sugar concentration up to 60%[17]; sugar concentration increased by 25% after a rise in temperature of 1° C at night[17]; secretion optimum at 22-25° C, but continues from 14-30° C and is increased with the addition of a full mineral fertilizer to the soil[17]; bees begin working sainfoin an hour before they begin foraging on *Trifolium hybridum* (alsike) or *Trifolium repens* (white clover)[72]; blooms for 10-14 days at an opportune time between fruit bloom (*Prunus, Pyrus*) and white clover (*T. repens*)[72] or between dandelion (*Taraxacum officinale*) and sweet clover (*Melilotus*)[35], and 2 weeks before alfalfa (*Medicago sativa*)[72]; seems more attractive to bees than *Trifolium repens* (white clover)[72].
Honey: yellow in colour; classed as water-white to light amber; granulates rapidly, finely, and to a solid consistency; flavour is less sweet than most other H; aroma faint, delicate[17]; probable source of famous French Gâtinais H[72].
Notes: cultivated forage crop; valuable pasturage plant that is only cut once; not long-lived in all areas; deep tap root and is drought resistant; pink flowers on a spike; see Table 18.

Opuntia polyacantha Haw.
prickly pear cactus, plains prickly pear cactus, Indian fig
per.   -37° to -29° C   Jun-Jul   20 cm   (N)P
Value for honey: N flow is very short (4-5 days)[72]; bees may collect red juices of the fruits[50]; other *Opuntia* are also attractive to bees[72].
Honey: reddish, classed as light amber; granulates to large crystals in clear liquid; heavy body and is described as stringy; fair to rank flavour[50,72].
Notes: large yellow flowers; useful for a hot dry position.

Origanum vulgare L.
wild marjoram, pot marjoram, origano, organy, marjolaine sauvage, origan
per.   -37° to -29° C   Jul-Aug   75 cm   N
Value for honey: HP4[16]; N sugar concentration of 76% has been recorded, making this N one of the most concentrated known[16]; H yield of 22 kg/colony/season reported where marjoram is a common weed[50]; blooms at the same time as fireweed (*Epilobium angustifolium*)[72].
Honey: is considered to improve the flavour of milder H's when blended with them[35]; a major source of H from Mt. Hymettus in Greece which is famous from classical times[16].
Notes: a cultivated culinary herb that also has medicinal uses; woolly aromatic leaves, purplish or white flowers; roadsides, old fields and thin woods, sw. Que. and s. Ont.

and beyond Can. limits[23].

Paeonia L.
peony
per.   Jun-Jul   45-180 cm   P[35]
Value for honey: single cvs. (only) are an excellent P source[35].
Notes: remarkably free of pests and diseases and also very hardy; will grow in any well-drained soil (slightly acid soil is best); does not like to be disturbed, and should not be transplanted more frequently than is necessary.

Papaver orientale L.
Oriental poppy, pavot de Tournefort
per.   -50° to -29° C   late May-Jun   60-120 cm   P
Value for pollen: P is highly attractive to bees (not unusual to see 3 or more bees gathering P at the same time in one blossom)[35]; P may have a narcotic effect on bees[72]; P from *P. somniferum* (opium poppy, pavot somnifère) has been implicated in serious mortality of foraging bees[7]; see Section 10.
Notes: among the longest lived poppies; should not be transplanted more than is strictly necessary.

Parthenocissus quinquefolia (L.) Planch.
(syn. *Vitis parthenocissus quinquefolia* (L.) Lamb.; *Ampelopsis quinquefolia* (L.) Michx.)
Virginia creeper, vigne vierge grimpante
per.   zone 3   Aug   less than 17 m (clinging vine)   N
Value for honey: H yield of 7-11 kg/colony/season[17]; highly attractive to bees[72].
Honey: slightly reddish in colour, classed as light amber; flavour distinctive[17].
Notes: rapidly growing vine with ornamental foliage and inconspicuous flowers; also a good ground cover for slopes and banks; woods and rocky banks, N. to se. Me., Vt., sw. Que., N.Y. and S. beyond Can. limits[23].

Penstemon barbatus (Cav.) Roth.
bearlip penstemon
per.   -50° to -29° C   Jun-Jul   90-120 cm   N[35]
Value for honey: valuable to help bridge the June gap in N flow.
Notes: loose panicles of bright red flowers.

Penstemon grandiflorus Nutt.
large flowered beard tongue
per.   -37° to -29° C   Jun   60-150 cm   N
Value for honey: similar to *P. barbatus* (bearlip penstemon)
Notes: thrives on a hot dry bank in full sun.

Physostegia virginiana (L.) Benth.
false dragonhead, obédience
per.   -50° to -29° C   Aug-Sep   90-120 cm   N(P?)
Value for honey: attractive to honeybees[72].
Notes: soft tubular pink flowers in close spike; moist, slightly acid soil; useful ornamental for borders and wild gardens; river banks, damp thickets and waste land, s. Que., s. N.B., N.E. and beyond Can. limits[23].

Polemonium caerulum L.
Jacob's ladder, Greek valerian, charity, échelle de Jacob, valériane Grecque
per.   -50° to -29° C   Jun-Jul   NP[36]
Value for honey: HP3/4[17].
Notes: attractive blue flowers with prominent yellow stamens; rich soil preferred; not native, but spread from cultivation to roadsides and waste land, e. Canada, N.E. and elsewhere[23].

Polygonum amplexicaule D. Don
mountain fleece
per.   -21° to -15° C   Jul   90 cm   NP[36]
Value for honey: HP4 (for *Polygonum*)[16].
Honey: light amber in colour; flavour is pronounced and unpleasant; very white cappings[16].
Notes: profuse white flowers.

Polygonum aubertii L. Henry
silver lace vine, silver fleece vine, China fleece vine
per.   zone 6   Aug-Sep   less than 3 m   N
Value for honey: HP4 (for *Polygonum*)[16]; "bees work this so eagerly that at times it sounds as if there may be a swarm of bees in the air"[72].
Honey: see *P. amplexicaule* (mountain fleece).
Notes: dense panicles of white fragrant flowers; excellent on chain link fences; very vigorous growth; not native, spread from cultivation to waste land[23].

Polygonum cuspidatum Siebold & Zucc.
(syn. *Reynoutria japonica* Houtt.)
Japanese knotweed, fleece flower, Mexican bamboo
per.   -37° to -29° C   Aug-Sep   1.8-2.4 m   N
Value for honey: HP4 (for *Polygonum*)[16]; said to "swarm" with bees[72].
Honey: see *P. amplexicaule* (mountain fleece)
Notes: an extremely vicious spreader and should be grown in waste land only; found in waste land and neglected gardens, rapidly spreading and becoming obnoxious, Nfld. to Ont. and S.[23]; Sakhalin knotweed (*P. sachalinense* Friedr. Schmidt ex Maxim.) also attracts bees in large numbers and is likewise a vicious spreader.

Pycnanthemum virginianum (L.) Pers.
basil, mountain mint, pycnanthème
per.   NP
Notes: unusually large amounts of bloom on each plant; needs no special care; native to parts of USA including centr. Me. to N.D. on gravelly shores, in meadows and dry to wet thickets etc.[23].

Ranunculus L.
buttercup, crowfoot, renoncule
per.   Jun-Aug   P
Value for pollen: P and plant juices contain highly toxic substance, protoanemonin[7]; nurse bees were observed to die when buttercup P was collected[7]; toxicity is retained in stored P for at least 3 years[7]; damage to bees is influenced by the prevalence of buttercup in pastures, competition from other sources and the weather[7]. See Section 10.

Notes: includes about 250 spp. of widely distributed herbs.

Rudbeckia laciniata L.
cut leaf coneflower, rudbeckie lacinée
per.   hardy   Jul-Sep   2.2-4 m   N[35]
Value for honey: sometimes freely worked for N[35]; see *Rudbeckia* .
Notes: Even single flowers provide only a small amount of N[35]; yellow flowers found growing wild from Que. to B.C., especially in rich low ground, w. Que., to Mont. S. to N.S., N.E. and beyond Can. range[23].

Salvia L. (see individual spp. for more detail)
salvia, sage, ramona, sauge
per. (bienn.)   Jun-Jul   N
Value for honey: HP4/6[16]; *S. mellifera* Greene (black sage) is not often cultivated, but is usually credited as the principal source of sage H in N. America[72]; *S. nemorosa* L. (wood sage) has a reported HP5[17] and occurs naturally in parts of the Okanagan and Fraser Valleys of B.C. Local opinion in B.C. holds that bees never collect anything from the common sage brush[a] in B.C.[b]
Honey: probably most of the many spp. of wild and cultivated sages give H with similar characteristics[72]; considered to be some of the finest H on the market[72]; classed as water-white; granulates extremely slowly (in some cases it has been known to stay liquid for years); heavy body; delightful flavour[16,72].
Notes: includes more than 750 spp. of herbs, subshrubs and shrubs usually growing in dry to stony sites; flowers mostly large and showy in spiked racemes or panicled whorls[23].

Salvia officinalis L.
garden sage, common sage, sauge, grand sauge, thé de France
per.   -37° to -29° C   Jun-Jul   20-70 cm   N(P?)
Value for honey: HP5/6[16]; N sugar concentration 30-60%[17].
Honey: light and pale yellow in colour; classed as amber to white; granulates slowly, like most *Salvia*; aromatic flavour[17];
Notes: much grown as a pot-herb or spice; commercially cultivated in the USA[51].

Salvia pratensis L.
meadow sage, meadow clary, sauge des près, sauge sauvage
per.   -37° to -29° C   60 cm   N[35]
Value for honey: see *Salvia*.
Notes: commonly cultivated; whorled racemes of (typically) bright violet blue or pink or white flowers; in cultivation flowers are extremely variable; naturalized in waste land, fields etc., Mass. and S. beyond Can. range[23].

Salvia × *superba* Stapf (: *S.* × *sylvestris* × *S. villicaulis* Borb.)
(syn. *Salvia virgata* Hort. not *Jacq.*; *S. virgata nemorosa* Hort.; *S. nemorosa* Mottet and auth., not L.)
per.   -23° to -21° C   mid Jun-late Aug   N[36]
Value for honey: HP4 (for *Salvia* × *superba* vars.)[16]; see *Salvia*.

---

[a]   The common sage brush may refer to an *Artemisia* sp.
[b]   McCutcheon, D. (1986). *Personal communication.*

Honey: see *Salvia*.
Notes: spikes of deep purple flowers with reddish bracts; known only in cultivation.

Satureja montana L.
winter savory, sarriette vivace, savourée
per.   -23° to -21° C   Jul-Oct   N
Value for honey: HP3 (for *Satureja*)[16]; very attractive to honeybees for N; may be more attractive than *S. hortensis* L. (summer savory, sarriette annuelle)[72].
Honey: similar to thyme (*Thymus*) H[16]; and is said to be greenish in colour[17].
Notes: flowers are pinkish white to purple; a shrubby culinary herb; *Clinopodium vulgare* L. (wild basil savory) is also attractive, but is less commonly cultivated[17].

Scabiosa caucasica Bieb.
Caucasian scabious
per.   -50° to -29° C   Jul-Sep   45-75 cm   N
Value for honey: yields well only in moist conditions; good N plant[35].
Notes: light blue flowers.

Scilla siberica Andr.
Siberian squill, scille de Sibérie
per. (bulb)   -50° to -29° C   Apr-May   15 cm   NP[36]
Notes: an excellent choice for naturalizing; will self-seed and eventually produce a carpet of deep blue blooms.

Scrophularia marilandica L.
Maryland figwort, Simpson honey plant, carpenter's square
per.   -29° to -23° C   Aug-Oct   0.9-3 m   N(P)
Value for honey: HP6 (for *Scrophularia* L.)[16]; excellent reputation as a N source that goes back at least a century (i.e. in the 1880's beekeepers planted figwort on a field scale)[72]; said to bloom for about 9 weeks when little else is blooming[72].
Honey: is light and of excellent quality; little aroma[16].
Notes: not usually grown as an ornamental; rich woods and thickets in sw. Me., sw. Que.[23] and beyond Can. limits.

Sedum spectabile Boreau
showy stonecrop, showy sedum, orpine
per.   -37° to -29° C   Sep-Oct   30-45 cm   N
Value for honey: attracts bees to some extent[35].
Notes: showiest of all the sedums; flat flower clusters, composed of many rosy pink flowers; sun or light shade; rocky or dryish soils.

Sidalcea candida A. Gray
white checker mallow, prairies mallow
per.   -23° to -15° C   Jul-Aug   90-120 cm   N
Value for honey: very attractive to bees.
Notes: miniature hollyhock-like, white to pinkish flowers.

Solidago L.

goldenrod, verge d'or, verge d'orée

per.   -37° to -23° C   15-180 cm   N(P)

Value for honey: HP4/5[16]; different spp. and var. vary widely in their value to the beekeeper, and there is little reliable information on the subject[72]. The good reputation of goldenrods among beekeepers may be largely due to plant abundance rather than high N secretion of individual plants[72].

Species suited to moist conditions do not appear to be attractive to bees when grown on dry sites[94]; later flowering spp. are said to be more attractive[50] and this is supported by several P.E.I. beekeepers[a] ; *S. canadensis* (Canadian goldenrod) and *S. occidentalis* (western goldenrod) are two of the most important spp. to beekeepers in Canada[72].

Honey: is deep golden in colour; thick, and granulates very quickly; aroma is strong and usually very pleasant[72].

Notes: about 125 spp. belong to this genus and most are native to N. America and many are common in every province of Canada[94]; much more ornamental use could be made of this group in N. American gardens since the flowers are very attractive and late flowering[95]; plants may become weedy if the soil is too rich; more susceptible to frosts than *Aster* (aster)[72]; undoubtedly most goldenrods are significant N and P sources, but recorded information is scarce; goldenrod is often mistakenly considered a major cause of hayfever[95]; see Table 18.

Solidago canadensis L.

(syn. *Aster canadensis* Kuntze; *Doria canadensis* Lunell; *Solidago altissima* L.; *S. lepida* DC.; *Aster lepidus* Kuntze)[b]

Canadian goldenrod, Canada goldenrod, verge d'or du Canada

per.   -37° to -29° C   Jul-Aug   30-120 cm   NP

Value for honey: HP4/5 (for *Solidago* L.)[16]; N sugar concentration 31%[94]; honey sacs of collecting bees were noted to be large (Colo.)[94]; bees appear to work mainly on plants growing on moist soil and plants in dry sites were largely ignored[94]; see *Solidago*.

Honey: see *Solidago*.

Notes: extremely showy plants; deep yellow flower-heads compactly arranged on a branching one-sided terminal panicle; stems clustered or solitary, very variable with several regionally more or less differentiated types which have been considered to represent different spp.; typical spp. (i.e. *S. canadensis* L.) is found naturally in moist to dry thickets, roadsides, clearings, slopes, Nfld. to Man., S. to N.S., N.E. and etc.[23]; ssp. *elongata* (Nutt.) Keck. is found naturally in B.C.[95]; var. *gilvocanescens* Rybdl.[23] (yellowish-gray) is found in Sask., E. to Mich. and beyond[23]; var. *hageri* Fern.[23] is naturally found in sw. Que. to Mich., S. to w. N.E. and beyond Can. range[23]; see *Solidago*.

---

a   Cosgrave, L. (1985). *Personal communication.*

b   Cronquist, A. (1955). Vascular Plants of the Pacific Northwest, Part 5. *University of Washington Press, Seattle.* 343 pp.

Solidago occidentalis (Nutt.) T. & G.[65]
western goldenrod, verge d'or
per.   50-100 cm   NP
Value for honey: HP4/5 (for *Solidago* L.)[16]; N sugar concentration 33%[94]; an important secondary source of fall N (and P) in Alta. and B.C.[94]; see *Solidago*.
Honey: see *Solidago*.
Notes: yellow flowers crowded into small daisy-like heads; clusters of flowers are flat topped; grows in wet meadows, river-banks, lakes and sloughs, found naturally in B.C. and Alta.[65]; see *Solidago*.

Stachys byzantina C. Koch
(syn. *Stachys lanata* Jacq., not Crantz; *S. olympica* of auth. not Poir.)
lamb's ears, woolly betony, woolly groundwort, épaire, épairie laineuse
per.   -29° to -23° C   Jul-Oct   30-45 cm   N
Value for honey: HP5[16].
Notes: usually grown for its attractive foliage; pink to purplish flowers on stalks are long lasting; a much cultivated ornamental.

Taraxacum Wiggers and *Taraxacum officinale* Wiggers
dandelion, lion's tooth, blowballs, dumble d'or, pissenlit
per.   -37° to -29° C   May-Sep   NP
Value for honey: HP4[16]; H yield 9-95 kg/colony/season[17]; N sugar concentration 18-74%[17]; N secretion 1.0-1.9 mg/flower/day; but none if temperatures are greater than 24° C[17]; flowers close on cloudy days[17]; no other N source of greater value for honeybees during the short main bloom period, but most H consumed during brood rearing because majority of colonies are not strong enough to take advantage of early flow[17,72]; flowers intermittently from spring to autumn; may be more attractive to honeybees than apple (*Malus*) when it blooms in orchards[35]; provides a major H flow in late May and June in the Eastern Townships, Montreal and Ottawa Valleys of Que.[12]; useful for colony build up in B.C.[a] and Alta.
Value for pollen: P yield is 1.2 mg/flower/day with approximately 6.2 mg/load[17]; protein content is high[83]; under-represented[b] in H[17]; load is dull yellow to orange[17]; considered a very valuable source of P; may be more attractive to honeybees than apple P when it occurs in orchards[35].
Honey: intense yellow to pale yellow in colour[17]; classed as light amber to golden amber[17]; granulates rapidly to a coarse solid[16]; flavour is sharp[91], but can be mild[50]; aroma pronounced[16].
Notes: *Taraxacum* includes many cosmopolitan weeds native to Europe, Asia and N. America (*T. officinale* is one sp. introduced to N. America); bright yellow flower-heads up to 3 cm across; pernicious lawn weed, but is also grown as edible greens; flowers intermittently all summer after main bloom in late spring.
    The following spp. and var. are common in certain parts of Canada and may contribute significantly to the H flow in their own areas, but no specific information has been recorded: *T. officinale* var. *palustre* (Sm.) Blytt.[23] is found in damper soils, Nfld. and Que. to s. N.E.[23]; *T. phymatocarpum* Vahl.[23], Greenl. and Arct. America, limestone crevices, Pistolet Bay, Nfld. (Jul)[23]; *T. ceratophorum* (Ledeb.) DC.[23], Lab. to B.C., S.

---

[a] Mitchell, S. (1987). *Personal communication.*
[b] This means that the percentage of pollen grains of this species to the total pollen grains in honey will be characteristically low even if most of the honey surplus originates from this species.

on calcareous ledges and cliffs to w. Nfld., Gaspé and Rimouski Cos., Que. (Jun-Aug)[23]; *T. laurentianum* Fern.[23], calcareous ledges and meadows, shores and ledges, w. Nfld. and Mingan Ids. and Anticosti I., Que. (Jul-Aug)[23]; *T. ambigens* Fern.[23] on calcareous meadows, shores, ledges, se. Lab. to Ung. S. to Nfld. and Gaspé Pen., Que. (Jun-early Aug)[23]; and var. *fultius* Fern.[23], w. Nfld. and Sickshock Mts., Que.[23]; *T. latilobum* DC.[23], rich, often calcareous slopes and talus, Nfld. to w. Me., adv. to w.N.E.[23]; *T. dumetorum* Greene[23], meadows and calcareous ledges, B.C. and Alta., to Colo., Slate Ids., L. Sup., Ont.[23]; *T. lacerum* Greene[23], calcerous turf and gravel, nw. Nfld., and Gaspé Co., Que. (late Jun-early Aug)[23]; *T. erythrosperum* Andrz.[23], thin dry soil, s. Que. to s. B.C., S. to N.S., N.E. and S. (Apr-Jul)[23]; *T. lapponicum* Kihlm.[23], damp meadows, ledges and shores, Greenl. to Alaska, S. to Nfld., Gaspé Pen., Que., Ung. along Rocky Mts. (Jun-Aug)[23].

Thalictrum L.
meadow rue, pigamon
per.   some very hardy   Jun-Jul   P
Value for pollen: bees work this assiduously for P[35].
Notes: ornamental blue-green foliage and insignificant flowers; blossoms have no true petals.

Thymus serpyllum L.
mother of thyme, wild thyme, creeping thyme, lemon thyme, serpolet, thym sauvage
per.   -37° to -29° C   Jun-Aug   49 cm   N
Value for honey: HP4[16]; H yield of 57 kg/colony/season[17]; N sugar concentration 27-45%[17]; N flow is usually dependable[72]; said to begin blooming at the same time as the basswood (*Tilia*) N flow is ending[72]; excellent pasture for honeybees[72].
Honey: very light in colour; classed as light amber to amber; flavour and aroma strong and minty; very good quality, but not ideal for over-wintering[17,72]; the famous H from Mt Hymettus (Greece) is derived largely from the wild thyme of that region[72].
Notes: excellent, low evergreen ground-cover; rosy purple flowers in terminal heads; used in rockeries or between stepping stones; sometimes found naturalized along roadsides; much material offered incorrectly as *T. serpyllum* L.; old fields etc., Que. and in Ont., S. to N.S. and S. to s. N.E.[23].

Thymus vulgaris L.
garden thyme, common thyme, faligoule, frigoule, thym, pote
per.   -23° to -21° C   Jun   10-30 cm   N
Value for honey: HP6[16]; H yield of 19 kg/colony/season[17]; N sugar concentration 36-45%[17].
Honey: golden amber in colour; classed as amber to medium amber; granulates slowly; flavour fairly strong, minty and does not fade on storage[17].
Notes: culinary herb; pink flowers on upright spikes; used ornamentally as an edging plant or as ground cover between stones.

Trifolium L. (see individual spp. for more detail)
true clover, clover, trèfle
per. bienn. ann.   mid summer   NP
Value for honey: most of the true clovers seem attractive to bees, but 4 spp. make up the bulk of the acreage and are of primary importance to the beekeeper (*T. incarnatum*, crimson clover; *T. pratense*, red clover; *T. hybridum*, alsike clover; and *T. repens*, white

clover)[54]; alsike clover, white clover and red clover account for a major part of the H produced in Canada, but they are much less widely grown than they once were[54,78].
Notes: about 300 spp. of leguminous plants of wide distribution in temperate and subtropical regions; some spp. are valuable forage and cover crops, some are constituents of lawns and a few are grown as ornamentals; propagation by seed broadcast in early spring.

Trifolium hybridum L.
alsike clover, alsike, hybrid clover, trèfle hybride
per.   Jun-Jul (Oct)   less than 60 cm   NP
Value for honey: HP4[16]; H yield up to 90 kg/colony/season[17]; N sugar concentration 26-47%[17]; beekeepers describe rapid gains on this clover (e.g. 32 kg/colony in 4 days)[72]; one of the best H plants in Canada[72]; N supply is increased by relatively low nitrogen and relatively high boron in the soil[17]; N secretion is stopped by drought[17]; see *Trifolium*.
Value for pollen: 45% of P collected at Beaverlodge Research Sta., Alta. over a 2 yr. period from alsike clover[66].
Honey: light in colour; classed as white; granulates rapidly; heavy body; flavour mild and delicate[17].
Notes: common pasture mix with timothy and red top grasses; used in crop rotations as an ann. or bienn.; could be used in a mix for seeding along roadsides; often spreads naturally to roadsides and clearings from Nfld. to B.C., and S. to n. USA.[23]; originally named as a hybrid from an early misconception that it was derived as a hybrid cross, whereas it is actually a true sp. See *Trifolium*.

Trifolium pratense L.
red clover, trèfle commun, Trèfle rouge
per.   -37° to -29° C   Jun-Aug   5-100 cm   N(D)P
Value for honey: HP3[16]; N sugar concentration up to 70[17]; plots treated with boron and ammonium molybdate showed a 53% increase in N secretion (and seed set) as well as an increase in N sugar concentration[17]; N secretion higher in hot dry summers[17]; first crop (of 2 or 3 crops in a season) often cut for hay during early bloom[17] and so may be of little value for N; continuing disagreement about the value of red clover as a H plant.
    The corolla of the red clover flower is longer than the honeybee's tongue, and only a small amount of N is accessible unless it is stolen by way of holes previously made near the base of the blossom. The holes are often made by insects collecting N or P, but not usually by the honeybee[54]. Late season cuts are more attractive to honeybees because the corollas of these plants are usually much shorter than the corollas of the first cut[54]; see *Trifolium* .
Value for pollen: load is dark green to dark reddish brown[17].
Honey: has a reddish or pink tinge; classed as water white to light amber; granulates rapidly; flavour mild and has nearly no aroma[17].
Notes: cultivated crop for hay, silage, pasture or green manure; fertile moist soil that is well-drained; average to heavy neutral loam and clay are recommended; lime deficiency is not tolerated[17]; roadsides, clearings and turf, Lab. to B.C., S. beyond Can. limits[23]; var. *sativum* (Mill.) Shreb.[23], cultivated red clover, is more common than *T. pratense*[23].

Trifolium repens L.

white clover, white Dutch clover, Dutch clover, trèfle blanc

per.   -37° to -29° C   Jun-Sep   10 cm   NP

Value for honey: HP3/4[16]; H yield 27-90 kg/colony/season[17]; N sugar concentration up to 64%[17]; N flow heaviest in season following a year of excessive rain[72]; N secretion best if soil is moist, air temperatures high, but less than 24°C and night temperatures less than 18°C, and on limestone soils[35]; in B.C. *T. repens* is claimed not to yield unless temperatures are 24°C or over[a]; N secretion is low on muck, sandy or lime-deficient soil and during periods of high humidity[17]; N sugar concentration was increased with the addition of boron to the soil[54] (application rate not indicated).

N flow starts 10 days after the flowering begins and ceases toward the end of the blooming period[35]; valuable to the beekeeper because it supplies a small amount of N over a long period[94].

Intermediate and small type cvs. are considered to be more attractive to honeybees than the larger forms. For instance, Ladino clover (*T. repens* L. Forma *lodigense* Hort. ex Gams) a large form of white clover, is apparently a poor N and P source in some areas[54,94]; see *Trifolium*.

Honey: bright yellow and clear; classed as extra-light amber to white; granulates rapidly with a slightly coarse grain (some observers disagree and contend that it granulates slowly with a fine smooth grain); flavour is mild and sweet[17].

Notes: a crop plant that maintains itself in a mix with Kentucky blue grass, unlike most spp. which are easily crowded out; may be suitable for inclusion in a roadside mixture; sometimes grown as an ann.; grasslands, roadsides and open pastures and woods throughout Canada[23]; see *Trifolium*.

Trillium L.

wake robin, birthroot, trillium, trille

per.   Jun   10-45 cm   P

Value for pollen: minor importance where it is naturalized in woodlands and wild gardens.

Tulipa kaufmanniana Regel

water lily tulip

per. (bulb)   -29° to -23° C   Apr-Jun   15-25 cm   (N)P

Value for honey: N sugar concentration of most tulips is low (approx. 5%)[8], but it may become more concentrated through evaporation in the open exposed blossom of the tulip; concentrated tulip N has been implicated in an occasional report of bee mortality, due to a large proportion of toxic sugars[8]; see Section 10.

Value for pollen: is sometimes attractive to bees for P; other single tulip spp. are also visited.

Valeriana officinalis L.

(syn. *Valeriana excelsa* Poir.)

garden heliotrope, common valerian, corn salad, lamb's lettuce, heal all, herbe à la femme meurtrie, valériane, herbe aux chats

per.   -37° to -29° C   Jul-early Aug   120 cm   N[36]

Notes: commercially grown for its roots which are used medicinally (dried rhizomes yield valerian); clusters of whitish-pink to lavender flowers with a weedy appearance;

---

a   McCutcheon, D. (1986). *Personal communication.*

waste land, roadsides, and old fields, N.E. and s. and w. USA.[23].

Verbesina alternifolia (L.) Britt. ex C. Mohr
(syn. *Actinomeris squarrosa* Nutt.; *A. alternifolia* (L.) DC.)
yellow ironweed, crownbeard, wingstem, golden honey plant, verbésine
per.   -29° to -23° C   Aug-Sep   1.5-2.4 m   N(P?)
Value for honey: freely worked by honeybees, and has a very good reputation as a N source in Canada[35].
Notes: coarse plant that is best grown in a "wild" garden or used for naturalizing; sun or part shade; needs rich soil; yellow flowers that form globose heads when mature; rich thickets and borders of woods, N. to N.Y. and s. Ont., also beyond Can. limits[23].

Veronica spicata L.
(syn. *Valerian australis* Schrad.)
spike speedwell, veronique en épi
per.   -37° to -29° C   Jul-Oct   60-90 cm   NP[36]
Value for honey: no bee garden should be without this plant[35]; never fails to attract bees[35].
Notes: blue to pink flowers borne in dense racemes; very neat appearance; popular garden ornamental; roadsides and rocky banks, Que. and n.N.Y.; naturalized[23].

Vitis vulpina L.
frost grape, winter grape, chicken grape, raisin, vigne
per.   zone 2   climbing vine   Jul   N
Notes: fragrant white flowers and edible fruit; river banks, bottomlands and rich thickets in parts of USA., N. to se. N.Y.[23]; see Table 18.

5.3 Trees and Shrubs

Abies alba Mill.
(syn. *Abies pectinata* (Lam.) DC.)
silver fir, sapin, argenté, sapin commun
tree   -29° to -23° C   less than 50 m   D
Value for honey: average honeydew H yield 40-45 kg/colony/season, maximum HP 96 kg/ha in parts of Europe (insect not specified)[7]; flow dependent on population size of aphid or scale insects feeding on sap[18]; value for H in Canada not documented; see Section 9.
Honey: (from insect, *Cinara pectinatae*) black or brown with a greenish tinge; also white to very dark green; granulates very rapidly; flavour of treacle; very sweet; mild to resinous[17]; see Section 9.
Notes: evergreen with pyramidal outline; native to temperate Europe; not often grown ornamentally in N. America; needs deep moist soil, cool moist climate; will not thrive where the summers are hot and dry; twig aphids may kill new growth[95]; to prevent the spread of balsam woolly aphid in B.C., planting and distribution of any spp. or cv. of *Abies* is prohibited in this province.

Abies concolor (Gord.) Lindl. ex Heldebr.
white fir, Colorado fir, sapin blanc
tree   zone 4   18-30 m   D(P?)
Value for honey: a notable source of honeydew H in Calif.[90]; probably a source in Ore.[79]; flow occurs earlier than that from *Calocedrus decurrens* (incense cedar)[90]; value is not well documented in Canada; see Section 9.
Value for pollen: reported that bees collect large loads of this light weight, wind borne P[90]; very low protein content[83].
Honey: may be similar to *Abies alba* (silver fir); see Section 9.
Notes: native to the Rocky Mts. and vicinity; one of the best firs for landscape use, and withstands city growing-conditions better than other firs; resistant to heat and drought; to prevent the spread of balsam woolly aphid in B.C., planting and distribution of any spp. or cv. of *Abies* is prohibited in this province.

Acer L. (see individual spp. for more detail)
maples, érable
trees and shrubby trees   Mar-late May   N(D occ., P)
Value for honey: HP4/5[16]; considered an important to major early N source in many parts of Canada; is likely that there would be a significant maple H surplus if colonies were stronger at this time of year. H is chiefly used by colonies for building up their populations in early spring.
      The following are known to have a good reputation among beekeepers and their order of bloom is given to make selections for continuous bloom easier[17,35,36,72,95].

Mar:	Acer saccharinum (silver maple, érable argentée)
early Apr:	A. rubrum (red or swamp maple, plaine rouge)
mid Apr:	A. negundo (box elder, érable à giguere)
late Apr:	A. circinatum (vine maple)
	A. palmatum (Japanese maple)
early May:	A. campestre (hedge maple, érable champêtre, petit érable)
	A. pseudoplatanus (sycamore maple, sycamore)
	A. saccharum (sugar maple, érable à sucre)
late May:	*A. ginnala (Amur maple)
Jun:	*A. pensylvanicum (striped maple, bois barré)

* These do not have a wide reputation as N sources.

Honey: pale amber and sometimes greenish in colour; slow, fine granulation; unremarkable flavour and aroma[16].

Notes: a large group of native and introduced trees; used extensively for lawn, park and street plantings; strong, free growing, but shallow rooted; grown also to provide wood for furniture; sap of one spp. (A. saccharum) yields well-known maple syrup and sugar; see Table 18.

Acer campestre L.
hedge maple, érable champêtre

| tree-like shrub | zone 5b | early May | 6-9 m | N(P) |

Value for honey: HP6[16]; secretes N freely; see *Acer*.
Honey: see *Acer*.
Notes: blooms for about 10 days; useful for screening purposes, but not commonly grown; not hardy at Ottawa; hardly naturalized (native to Europe).

Acer circinatum Pursh
vine maple, érable à feuilles rondes

| shrub-like tree | zone 6 | late Apr | 3-9 m | N(P) |

Value for honey: H yield in an exceptional year 22.6 kg/colony/season[17]; N sugar concentration 42%[17]; a major source of H in some areas (e.g. B.C.)[72]; see *Acer*.
Honey: amber with a distinctive flavour[17].
Notes: one of the most ornamental of maples in flower; excellent specimen tree; foliage turns orange to red in autumn; B.C. to n. Calif.[23].

Acer ginnala Maxim.
Amur maple

| tree-like shrub | zone 2 | late May-early Jun | 6-9 m | N(P) |

Value for honey: potentially more surplus H from this than from most maples because it blooms later; see *Acer*.
Honey: see *Acer*.
Notes: fragrant whitish flowers; vivid scarlet autumn colour; few pests[95]; locally established from Me. to Ct.[23].

Acer macrophyllum Pursh
Pacific maple, big leaf maple, Oregon maple, canyon maple, érable à grandes feuilles

| tree | zone 8 | early May | 15-21 m | N(P) |

Value for honey: N sugar concentration 52%[17]; a major H source in B.C. (but flow is often cut short by rain)[17]; see *Acer*.
Honey: see *Acer*.

Notes: yellow fragrant flowers; leaves turn to a bright orange or yellow in the fall; excellent shade tree; found naturally from se. Alaska to Calif.[23].

Acer negundo L.
Manitoba maple, box elder, érable à giguere
tree   zone 2   mid Apr   9-15 m   (N,D?)P
Value for honey and pollen: honeydew may be secreted by aphids feeding on leaves[72]; a noted source of early fresh P, and some N in Sask[53]; see *Acer*.
Notes: hardy, drought resistant; grows well in prairie provinces and is used there for shelter belt planting; naturally occurs along river banks, N. to w. N.E., N.Y. and s. Ont.; much cultivated and naturalized E. to Maritimes and e. Que[23].

Acer palmatum Thunb.
Japanese maple, érable du Japon
shrub-like tree   zone 6   mid May   3-6 m   N(P)
Value for honey: see *Acer*.
Honey: see *Acer*.
Notes: many cvs., some with brilliantly coloured foliage and finely divided leaves; very ornamental.

Acer platanoides L.
Norway maple, plaine, plane, faux sycamore
tree   zone 5   late Apr   15-23 m   N(DP)
Value for honey: HP4[16]; N sugar concentration 30-50%; honeydew also collected[17]; see *Acer* and Section 9.
Value for pollen: colour of load is dark green to brown[17].
Honey: see *Acer* and Section 9.
Notes: masses of small yellow flowers; foliage yellow in the fall; withstands city conditions; much planted; seedlings survive in hedgerows, roadside thickets etc.[23].

Acer pseudoplatanus L.
sycamore, sycamore maple, mock plane, grand érable, faux platane
tree   zone 5b   early May   12-21 m   N(DP)
Value for honey: HP4[17]; N sugar concentration 47%[17]; 2-3 week N flow[17] (long for maple); honeydew may also be collected[17]; see *Acer*, and Section 9.
Honey: light amber in colour and may have greenish tinge; slow to granulate[17]; see Section 9.
Notes: flowers borne in long panicles; grows well on exposed sites (inc. seashore areas) where other spp. might fail; dry to well-drained site preferred[17]; much planted; seedlings freely establishing in fence-rows, roadsides etc.[23].

Acer rubrum L.
swamp maple, red maple, scarlet maple, soft maple, plaine, plaine rouge
tree   zone 3   early Apr   18 m   N(P)
Value for honey: an important N plant in e. Canada when weather is conducive to N flow[72]; has been recommended for roadside planting in N.J. to increase bee forage[62]; see *Acer*.
Honey: see *Acer*.
Notes: flowers are bright red in spring and so are the leaves in autumn; fast growing and weak-wooded[95]; when in flower, a massed group of trees is like a scarlet haze;

seeds itself quickly in open areas, especially if poorly drained, so planting is not required when using as a ground cover[62]; few equals in poorly drained sites for retarding water run-off of surface water[62]; Nfld. and Gaspé Pen., Que. to Man., and S. beyond Can. limits[23].

Acer saccharinum L.
soft maple, silver maple, river maple, plaine blanche
Value for honey: an important source in Ont.[72]; H yield of 0.5-5 kg/colony/day reported[72]; see *Acer.*
Honey: see *Acer.*
Notes: much planted, but not a good permanent tree because it is weak-wooded; wide spreading habit; river banks and bottomlands, N.B. to Ont., and S. beyond Can. limits[23].

Aesculus L. (see individual spp. for more detail)
chestnut, buckeye, marronier
trees and shrubs   late May-mid Jun   NP(D)
Value for honey and pollen: all *Aesculus* contain toxic saponins, but they are rarely gathered in the N, P or D in doses lethal to honeybees[7]; the saponin content may vary widely between individual plants and in different years[7]; *A. californica* (California buckeye) is probably the most toxic plant known to N. American beekeeping and *A. hippocastanum* (common horse chestnut, marronier) has been the reputed cause of bee mortality in Europe[7]; see Section 10.
Honey: reported as both light and dark; dense, and of medium quality[16].

Aesculus × carnea Hayne (: *A. hippocastanum × A. parvia*)
red horse chestnut, châtaignier de cheval rouge
tree   zone 5b   late May-mid Jun   12-18 m   NP
Value for honey and pollen: HP4[16]; flowers about 2 weeks later than *A. hippocastanum* (common horse chestnut), at a useful time when there may be little else available[35]; see *Aesculus.*
Honey: see *Aesculus.*
Notes: pink to red flower clusters 25 cm tall, are its chief ornamental asset.

Aesculus glabra Willd.
Ohio buckeye, pavier glabre
tree   zone 2b   mid-May   6-18 m   NP
Value for honey: H yield 14 kg/colony/season reported for strong colonies[72]; see *Aesculus.*
Value for pollen: major (more than 10% of the P collected) source at Beaverlodge, Alta. in 1977, but apparently not visited for P in 1978[66].
Honey: very thick and the colour of basswood (*Tilia*) H, but not equal to the latter in quality[72].
Notes: greenish-yellow flowers; foliage turns orange in the fall.

Aesculus hippocastanum L.
common horse chestnut, horse chestnut, European horse chestnut marronier, châtaignier de cheval
tree   zone 5b   mid May-mid Jun   23 m   NP
Value for honey and pollen: HP2/3[17]; N sugar concentration 30-76%[17]; N flow lasts

about one month, but flowers may remain longer[35]; a toxic saponin is found in the sap, N and P which is poisonous to bees giving symptoms called "addled brood" or "May disease"[64]; poisoning occurs only when trees are intensely visited, as they are during a dry year[64]; poisoning from common horse chestnut has rarely been a problem in N. America although it has frequently been reported to be a problem in Europe[7]; see *Aesculus*.

Honey: almost colourless; granulates rapidly[17].

Notes: grows vigorously, 60 cm per year[17]; introduced and often self-seeds from cultivated trees.

Ailanthus altissima (Mill.) Swingle
tree of heaven, varnish, copal tree, ailante glanduleux, vernis du Japon
tree   zone 6   mid Jun-late Jun   9-12 m   N(P)
Value for honey: surplus H reported where sp. is common (e.g. Vienna, Austria and Paris, France)[35].

Honey: fresh H may have an unpleasant taste, but this improves with age, developing into a Muscatel-like flavour; pale greenish-brown in colour; crystallizes to a fine grain in three months[35].

Notes: grows better under city conditions than any other native or introduced tree known[95]; yellow flowers in small clusters; sexes separate; female trees are usually grown because the male flowers have an objectionable odour (like male cats)[35,95]; naturalized widely in Mass. to s. Ont. and S.[23].

Alnus B. Ehrh.
alder, aune
shrubs and tree-like shrubs   Mar-Apr   P
Value for pollen: produces a copious amount of P in the early spring which is often eagerly collected by honeybees if the weather is favourable[72]; considered a valuable source by beekeepers; P is poor in protein content and is not particularly good for brood rearing[83].

Notes: several of these legumes are native to Canada; coarse foliage; will grow in wet, boggy or shady locations; see Table 18.

Alnus incana (L.) Moench
speckled alder, European white alder, white alder, aune blanchette, aune de montagne
shrub   zone 4   early Apr   4-15 m   P
Value for pollen: P abundant; *A. rugosa*, native to e. Canada, offers P at the same time; see *Alnus*.

Notes: see *Alnus*.

Amelanchier alnifolia Nutt.
Saskatoon, service berry, June berry, shadbush, poirier, petites poires
shrub   zone 1   early May   1.8-3 m   NP
Value for honey: recognized as a valuable N plant in n. Man., Sask., and Alta., but not as a source of surplus H[57,72]; *A. canadensis* (Canadian June berry) is also attractive to bees[35]; other *Amelanchier* are likely to be some importance for N and P, but few records of observations.

Notes: masses of white flowers that are short lived, followed by blue-black edible fruits[81]; tolerant of a low level of sunshine; native habitat includes, thickets, borders of woods and banks of streams from w. Ont. to Yukon, also nw. Colo. and Ore.[23].

Amorpha canescens Pursh
leadplant
shrub   zone 2b   20-30 cm   NP[42]
Notes: dense, dwarf bush; small dainty foliage; highly desirable for foundation planting; dry sandy soils in the prairies and hills, w. Sask. to Mich. and S.[23].

Amorpha fruticosa L.
false indigo, river locust, bastard indigo, indigo bush, amorphe arbustif, faux indigo, indigo bâtard
shrub   zone 2b   late Jun-early Jul   6 m   NP
Value for honey: blooms between fruit and clover[72].
Notes: large vigorous shrub, particularly suitable on the prairies; flowers purplish blue.

Andromeda L.
bog rosemary, andromeda, andromedie, andromède
shrub   NP
Value for honey and pollen: some of these are suspected of producing a H (and maybe N) poisonous to humans[6].
Notes: grown in rock gardens and borders; very hardy; *A. glaucophylla* Link., bog rosemary, found in peats, bogs etc., sw. Greenl. and Lab. to e. Man., S. to Nfld., N.S., N.E. etc.[23].

Aralia elata (Miq.) Seem.
Japanese angelica tree, Japanese angelica, angélique en arbre du Japon
shrub   zone 5   Aug   2-8 m   N(P?)
Value for honey: similar in value to *A. spinosa* (Hercules' club).
Notes: very slow growing, striking ornamental with large bunches of white flowers.

Aralia spinosa L.
Hercules' club, angelica tree, devil's walking cane, prickly ash, angélique en arbre d'Amerique, gourdin d'Hercule
shrub   zone 6   Aug   3-10 m   N(P?)
Value for honey: very attractive for N; N clearly visible in the blossoms[35]; rarely a source of surplus H[72]; said to bloom at the same time as buckwheat (in Penn.)[72]; has been recommended for planting along roadsides (N.J.) to increase bee forage[62].
Honey: light in colour; good body and fair flavour[72].
Notes: striking ornamental with large trusses of small white flowers; suckers, and will form a thicket[62]; bluffs, rich woods and river banks, escaped from cultivation N. to s.N.E., centr. N.Y. and Mich.[23].

Arbutus menziesii Pursh
Pacific madrone, madrona, arbutus, arbute, arbousier
tree   zone 2   May   9-14 m   NP
Value for honey: a particular source of H in B.C.[73]; visited by bees in large numbers[72].
Honey: is light golden amber; very heavy body; good flavour[72].
Notes: lily-of-the-valley-like flowers; evergreen; orange coloured berries; found from B.C. to Calif.[23].

Arctostaphylos columbiana Piper
woolly manzanita, hairy manzanita
shrub   -15° to -12° C   Mar-May   N(P?)
Value for honey: an important source, said to encourage brood rearing[72]; on warm days, N can be shaken from the blossoms[72].
Notes: flowers white or pink, borne in panicles; leaves woolly on the under-surface; coastal B.C. to Calif.[23].

Arctostaphylos manzanita Parry
common manzanita, Parry manzanita
shrub   -15° to-12° C   Apr-May   30 cm   N(P?)
Value for honey: yields N freely[72]; an important source on the w. coast; blooms for about 6-8 weeks[72].
Honey: white or light amber, with a delicious flavour[72]; n. Calif. to s. B.C.[23].

Arctostaphylos uva-ursi (L.) K. Spreng.
bearberry, kinikinnik, common bearberry, mealberry, hog cranberry, bear's grape, bousserole, raisin d'ours, thé de brousse
ground cover   zone 1   May-Jun   N(P?)
Value for honey: reports of an average yield of 18 kg/colony/season (this in combination with *A. manzanita*)[50]; an important source in B.C.
Honey: light amber; good, but slightly bitter flavour[50].
Notes: native to the coldest parts of Canada; ideal on poor soils where few other plants will grow; pink flowers followed by berries[81]; Arct. reg., S. on exposed rock and sands to w. Nfld., Que.; var. *adenotricha* Fern. & Macbr.[23] found in Côte Nord, Que., to B.C., S. to Mont.[23].

Berberis L.
barberry, épine vignette, vinetier
shrubs   spring   NP
Value for honey and pollen: most, if not all the barberries are visited by bees, and some are important, native sources[72].

Several barberries act as alternate host for the black stem rust of wheat and are being eradicated from wild areas and gardens[81]. Planting of any of the deciduous barberries cannot be recommended for this reason and only those evergreen barberries known to be immune to the disease and attractive to honeybees are mentioned here. *Mahonia aquifolium*, Oregon grape, is sometimes incorrectly referred to as a *Berberis*.

Berberis darwinii Hook.
Darwin's barberry, vinetier de Darwin
shrub   zone 7   Mar-May   1.2-2.0 m   NP
Value for honey: observed to be one of the most attractive of the barberries to honeybees[72].
Notes: yellow flowers in pendant racemes; suitable only for very mild regions.

Berberis × *stenophylla* Lindl. (: *B. darwinii* × *B. empetrifolia* Lam.)
rosemary barberry
shrub   zone 6b   Mar-May(?)   2.4 m   NP
Value for honey: observed to be one of the most attractive barberries to honeybees[35].
Notes: a tall graceful specimen plant; evergreen.

Buddleia globosa Hope.
globe butterfly bush
shrub   -15o to -12° C   Jun   4.5 m   N(P?)
Value for honey: very attractive to honeybees for N; other *Buddleia* are adapted to
butterfly pollination and are of no use to the honeybee since the corollas are too long
(e.g. *B. davidii*, summer lilac or orange-eye butterfly bush)[35]; a good choice for those
interested in "bee-watching", since many of the small flowers that constitute the spheres
of bloom are worked by bees in quick succession while bees remain nearly still[35].
Notes: bright, showy fragrant flowers; cultivated on the Pacific coast.

Callicarpa bodinieri Lév. cv. *giraldii* (Hesse ex Rehd.) Rehd.
(syn. *Callicarpa giraldiana* Hesse)
Bodinier beautybush, Bodinier beautyberry
shrub   zone 6b   Jul   120 cm   N
Value for honey: favoured by bees[35].
Notes: lilac coloured flowers; should be clumped for best effect.

Calluna vulgaris (L.) Hull
heather, ling, Scotch heather, bruyère, bruyère commune, bucane
shrub   zone 6   Aug-Nov   4 cm   NP
Value for honey: HP4[16]; N sugar concentration 25-50%[17]; greatest N secretion on 1-2
year shoots, following burning, and on newly opened flowers[17]; may be locally
important in parts of the Maritimes, especially N.S.[72]. See Table 9.
Honey: light to dark reddish brown in colour; granulates slowly and forms large
crystals; thixotropic (due to protein content) which makes it difficult to extract fresh
combs; unsuitable as sole winter food for bees[17]; a strong market for this H.
        Phyllodoce Salisb. (false heather, mountain heather) is a locally important source
of some of the so-called "heather" H in B.C.[72].
Notes: acid, damp soil of low fertility best[17]; common in local areas of N.B., N.S. and
Nfld. where it has naturalized and spread after introduction; many cvs.; prune well in
the spring for maximum blossom.

Calocedrus decurrens (Torr.) Florin
California incense cedar, white cedar, cedre blanc
tree   zone 8   12-27 m   D
Value for honey: honeydew H from scale insect, *Xylococcus macrocarpa*, abundant at
times, from early summer onwards[17]; 45-136 kg/colony/year[72]; honeydew H in
commercial quantity in Ore. and Calif.[72]; see Section 9.
Honey: classed as amber to dark amber[17]; noted to be white in colour in Jul (Calif.)[17];
very gummy; granulation very slow; flavour bland; heavy body; water content low, 12-
15% (sp. of insect not specified)[17]; see Section 9.
Notes: can be found in association with *Abies concolor* (white fir)[72]; suitable for formal
plantings; scale-like lustrous foliage; good soil and ample moisture.

Caragana aborescens Lam.
Siberian peashrub, Siberian peabush, Siberian peatree, caragana arborescent
shrub   zone 2   mid May-early Jun   3.6-5.5 m   NP
Value for honey: HP6[16]; a source of surplus on the n. prairies[72]; said to give a 2 week
flow in Morden, Man.[72]; *C. pygmaea* (pygmy peashrub) is visited by honeybees to
some extent, but the shape of the blossoms is more ideally suited to bumblebees[35] and

makes it difficult for the honeybee to obtain N[94].
Honey: light in colour and of fine quality.
Notes: coarse, yellow pea-like flowers; drought resistant; grown for a windbreak on the prairies and sometimes in gardens; introduced.

Catalpa bignonioides Walt.
common catalpa, southern catalpa, Indian bean, catalpe commun
tree   zone 6   late Jun   12 m   NP
Value for honey: N secreted before, after and during the blooming period by extra-floral nectaries on the under-surfaces of leaves[35]; bees collect N from flowers by crawling right into the bell-shaped blossoms[35]; several other catalpas are not attractive to bees[35]; cultivated and frequently escaped, N. to s. N.E., N.Y.[23].

Ceanothus L. (see individual spp. for more detail)
buckbrush, redroot
shrubs   mid summer   NP
Value for honey: several native N. American spp. are considered very good N sources, but are not usually cultivated, these include: *C. sanguineus* (wild lilac), *C. prostratus*, *C. velutinus* (snowbrush, sticky laurel); the latter blooms in early July, yielding a surplus of 18 kg/colony/season of light amber H[72].
Notes: evergreen and deciduous flowering shrubs noted for their profuse display of blue, pink and white flowers; thrive in light, well-drained soil in sunny locations.

Ceanothus americanus L.
redroot, New Jersey tea, wild snowball, mountain sweet, céanothe
shrub   zone 4b   mid Jun-end Jul   90 cm   NP[20]
Honey: light amber and of good flavour[50].
Notes: white flowers in upright clusters; dry to open woods and gravelly or rocky banks; s. USA. to s. Que., S. to Ont., and s. Man.; leaves considered to be one of the best substitutes for tea during the American revolution[23].

Cephalanthus occidentalis L.
button bush, bois noir
shrub   zone 4b   late Jul-mid Aug(?)   2.4-3.6 m   N(P?)
Value for honey: surplus H reported (e.g. Fla. and Vt.)[72]; corollas may be too deep for the honeybee to consistently reach all the N[72]; in dry years, bloom is reduced[62]; recommended for roadside planting to increase bee forage (i.e. N.J.)[62].
Honey: light amber in colour; heavy body; flavour is mild[72].
Notes: native to e. Canada[81]; will thrive in areas flooded for most of the year[62] and grows well in groupings of *Clethra alnifolia* (summersweet) and *Acer rubrum* (red maple)[62]; roots readily by self-layering and needs full sun[62]; profuse cream-coloured globular flower clusters[62]; blooms when few other shrubs flower[81]; swamps, pond borders and stream margins, N. to w. N.S., sw. N.B., N.E., sw. Que., s. Ont.[23].

Cercis canadensis L.
redbud, eastern redbud, Judas tree, bouton rouge, gainier
shrub-like tree   zone 6   Apr-May(?)   3-9 m   N(P?)
Value for honey: most of the H collected is used in brood rearing[50]; usually blooms with fruit trees (i.e. *Prunus*)[72].
Honey: light coloured; well flavoured[62].

Notes: rosy pink flowers borne along the stems before the leaves appear; rich woods and ravines, se. N.Y., s. Ont., s. Mich., and s. Wisc.[23].

Chaenomeles speciosa (Sweet) Nakai
flowering quince, Japanese quince, cognassier
shrub   zone 5b   early May-mid Jun   12 cm   NP[36]
Value for honey: freely visited for N, but only if the weather is suitably warm[35]; *C. japonica* (Japanese quince) has been recommended for roadside planting to increase N production (i.e. N.J.)[62]; a valuable early source[35].
Value for pollen: visited chiefly for P[35].
Notes: many cvs. available (be sure to select only singles for bee forage); best planted in full sun against a wall, or can be grown as a hedge.

Cladastrus lutea (Michx. f.) C. Koch
(syn. *Cladastrus tinctoria* Raf.)
American yellow wood, Kentucky yellow wood, virgilia, virgilier à bois jaune, cladrastre à bois jaune
tree   zone 4b   early Jun   9-12 m   N(P?)
Value for honey: highly attractive source of N[72]; unfortunately, does not flower regularly[35], often only every third year[95].
Honey: light amber in colour, with a strong and distinctive flavour[50,72].
Notes: excellent ornamental for flowers and foliage; large pendulous clusters of white flowers and orange to yellow fall foliage; natural range limited to se. USA.[72]; but cultivated and often spreading N. to Mass.[23].

Clethra alnifolia L.
summersweet, sweet pepperbush, cléthra
shrub   zone 5   late Jul-late Sep   1-3 m   NP
Value for honey: H yield 68-77 kg/colony/season in an exceptional year[72]; H yield 34 kg/colony/season also noted[50]; N sugar concentration 20-43%[17]; H flow can be very heavy and is usually continuous for several weeks[62]; recommended for planting along roadsides in order to increase bee forage (i.e. N.J.)[62].
Honey: is white, may be tinged with yellow; thick; aroma like the flowers[17]; H among the finest produced[62].
Notes: creamy, waxy flowers borne in dense cluster; spicy fragrance[62]; prefers sandy, acid soils and sites near water[62]; found naturally with *Vaccinium* (blueberries) and *Acer rubrum* (red maple)[62]; best grown in clumps; swamps, damp thickets and sandy woods, N. to s. Me., s. N.H., Mass., s. N.Y.[23].

Colutea arborescens L.
bladder senna, baguenaudier, senné bâtard
shrub   zone 5   late May-early Sep   1.2-1.5 m   N(P)
Value for honey: not a wide reputation as a bee plant, but an attractive N source[35]; N is collected from the side of the blossom[35].
Notes: vigorous and self-seeds; could be considered for roadside or waste land plantings.

Cornus alba L.
Tartarian dogwood, Tatarian dogwood, cornouiller blanc
shrub   zone 2   early-late Jun   2.4-3.0 m   NP[36]
Value for honey: H yield 4.5-6.7 kg/colony/season (other *Cornus* may have contributed to this total)[50].
Honey: light golden in colour; flavour is mild.
Notes: coarse in appearance, but lovely vivid red stems in winter; useful for screening.

Cornus mas L.
Cornelian cherry, sorbet, cornouiller des bois, cornouiller mâle
shrub   zone 5b   early Apr-late Jun   3-7 m   NP[36]
Value for honey: N is occasionally reported[72]; not favoured by bees[72].
Notes: one of the earliest of the dogwoods to flower; specimen or shrub border.

Cornus sericea L.
(syn. *Cornus stolonifera* Michx.)
red osier dogwood, American dogwood, harts rouges
shrub   zone 1b   early Jun-late Jun   2.1 m   N[36]
Value for honey: very attractive to all nectar-collecting insects[62]; recommended for roadside planting to increase bee forage (i.e. N.J.)[62].
Notes: dense corymbs of minute, creamy blossoms; red stems in winter; bluish berries eaten by birds; can form dense thickets which will withstand the force of floods and effectively resist bank erosion[62]; shores and thickets, Nfld. and s.Lab. Pen. to Yuk., S. to N.S., N.E. etc[23].

Corylus americana Marsh.
American hazel, American filbert, noisetier, coudrier
shrub   zone 2b   Mar   3 m   P
Value for pollen: valuable only if no other source of P available[72]; P is poor in protein content[83].
Notes: mainly planted for its nuts as food for wildlife; catkins last for about a month; naturally found in thickets from centr. Me. to Sask., and S.[23].

Corylus avellana L.
European hazel, European filbert, noisetier commun, coudrier, avelinier
shrub   zone 5   Mar   4.5 m   (D?)P
Value for pollen: P eagerly collected if these are near over-wintered colonies[35]; protein content is poor[83]; P is light and powdery[35].
Honey: where this is grown as a crop, beekeepers report H from honeydew[72].
Notes: soft yellow cascades of catkins which have ornamental value; grown commercially for its nuts.

Cotoneaster Medic. (see individual spp. for more detail)
cotoneaster, rock cotoneaster
shrub   Jun-Jul   NP[36]
Value for honey: considered a first class bee plant and at least 2 dozen cultivated spp. of *Cotoneaster* are visited eagerly by bees[35]; few other plants are visited so persistently as cotoneaster[35]; may be more attractive than *Tilia* (basswood)[35]; the following are attractive to bees, but are not described here; *C. franchettii* (Franchett cotoneaster) in Aug; *C. conspicuous* (winter green cotoneaster) in Jul; *C. frigidus* (Himilayan

cotoneaster) in Jul-Aug; *C. simonsii* (Simon's cotoneaster) in Jul[35].
Notes: shrubs of very diverse habit, ranging from the prostrate ones to large shrubs; chiefly valued for their foliage and bright berries; full sun[81].

Cotoneaster horizontalis Decne.
rockspray cotoneaster
shrub   zone 6   mid Jun   90 cm   NP[36]
Value for honey: a favourite with bees[35]; see *Cotoneaster*.
Notes: very commonly grown; excellent as ground cover or as an espaliered specimen.

Cotoneaster microphyllus Wallich ex Lindl.
small-leaved cotoneaster
shrub   zone 7b   60 cm   NP[36]
Value for honey: see *Cotoneaster*.
Notes: suitable for rock gardens or foundation planting.

Cratageus L. (see individual spp. for more detail)
hawthorn, thorn apple, red haw, pommettes, cenellier, épine
tree or shrub   May-Jun   NP
Value for honey: HP2[16]; H yields of 4.5-7 kg/colony/season sometimes reported[50]; considered fickle yielders and not to be depended upon[35]; good years are usually infrequent, but when they do occur, the N flow is rapid[35]; recommended for planting along roadsides to increase bee forage[62].
Honey: light or dark amber in colour; flavour is fine, minty or nutty[16,50].
Notes: diverse in habit, but most have white spring flowers and apple-like fruit; all thrive in heavy, poorly drained soil[62], but prefer limestone or rich loamy soils.

Crataegus crus-galli L.
cockspur hawthorn, cockspur thorn, épine ergot de coq
large shrub or small tree   zone 2b   mid Jun   4-9 m   NP[36]
Value for honey: see *Crataegus*.
Honey: see *Crataegus*.
Notes: a good display of white flowers and berries in the fall; open ground, often in dry or rocky places, se. Canada[23]; see *Crataegus*.

Crataegus laevigata (Poir.) DC.
(syn. *Crataegus oxyacantha* of auth., not L.; *C. oxyacantha* Thuill.)
English hawthorn, white thorn, May tree, bois de Mai, abepine
large shrub   zone 6   late May   6 m   NP[36]
Value for honey: see *Crataegus*.
Honey: see *Crataegus*.
Notes: introduced and often an escaped; see *Crataegus*.

Crataegus punctata Jacq.
dotted hawthorn
tree   zone 5   late May-late Jun   NP
Value for honey: very attractive to honeybees and precedes the *Trifolium* (true clover) bloom[72]; considered a N source in Que., and Ont., where the bloom lasts about 3 weeks[72]; see *Crataegus*.
Honey: see *Crataegus*.

Notes: found naturally in open ground, thickets, pastures, e. Canada and N.E. and S.[23]; see *Crataegus*.

Cytisus L.
broom, cytise
shrub   May-Jun   (N)P
Value for honey: beekeepers sometimes contend that there is no N collected from brooms growing wild[35], but say that some cultivated brooms are noted to be N sources[35]. However, no spp. or var. actually does produce N[a].
Value for pollen: beekeepers claim that only P is collected from wild brooms[35]; an attractive P source[72].
Notes: free flowering every year, in poorest, dry soil.

Cytisus albus Hacq.
Portugese broom, cytise blanc
shrub   zone 7   late Jun   45 cm   NP[35]
Value for honey and pollen: see *Cytisus*.
Notes: see *Cytisus*.

Cytisus scoparius (L.) Link.
Scotch broom, genêt à balais
shrub   zone 7b   mid-May   180 cm   (N)P
Value for pollen: an important early source of P on Vancouver I., especially in the vicinity of Victoria[72]; also naturalized in parts of coastal N.S.
Notes: sandy roadsides, open woods and barrens, N.S., sw. Me. to w. N.Y. etc.[23]; see *Cytisus*.

Daphne cneorum L.
rose daphne, garland flower, petite thymelée, laureole odorante
shrub   zone 2b   mid-May   30 cm   NP
Value for honey and pollen: dependably attractive in warm weather[72]; is very toxic to mammals and although toxicity to bees is not proven, there is one account of honeybee poisoning by this N[7] (see Section 10).
Notes: a precise trailing evergreen shrub, useful as a ground cover; rose coloured flowers; one of the earliest of shrubs to bloom; naturalized in many parts of Canada, particularly N.S.; vivid red fruits are poisonous.

Deutzia × *kalmiiflora* Hort. Lemoine (: *D. parviflora* × *D. purpurascens*)
fuzzy deutzia
shrub   zone 6   late Jun   180 cm   NP
Value for honey: HP1 [16].
Notes: one of the last deutzias to flower.

Diervilla lonicera Mill.
dwarf bush honeysuckle, honeysuckle diervilla, herbe bleue
shrub   zone 2   early spring   90 cm   N(P?)
Value for honey: bees work this eagerly when weather permits[72]; most honeysuckle (*Lonicera*) vines or shrubs have a corolla that is too long for honeybees to reach all the

---

[a]   Adey, M. (1986). *Personal communication.*

N[50].
Notes: dry woods, clearings and rocky places, Nfld. to Man., S. to N.S., N.E.[23].

Diervilla sessifolia Buckl.
southern bush honeysuckle
shrub   zone 4   late Jun   120 cm   N
Notes: masses of sulphur yellow flowers; spreads by underground stolons; may be suitable for erosion control or holding banks[95]; upland woods of S. centr. USA.[23].

Diospyros virginiana L.
common persimmon, American persimmon, possum apple, possumwood, plaqueminier
commun
tree   zone 7   Jun   9-15 m   N(P)
Value for honey: H yield 14-32 kg/colony/season[17]; N flow is brief (2 weeks)[17].
Honey: is light amber to amber in colour[17].
Notes: white flowers; edible fruits; long-lived, spreads by stolons and grows in poor soil; dry woods and fields se. USA. to se. N.Y.[23].

Elaeagnus angustifolia L.
Russian olive, oleaster, wild olive, silverberry, olivier de Boheme, arbre d'argent
shrub   zone 2b   early Jun   6 m   N
Value for honey: HP3[17]; N sugar concentration 35-42%[17]; visited all day for N[72].
Notes: inconspicuous fragrant yellow flowers; dry site; introduced.

Elsholtzia stauntonii Benth.
heather mint, Staunton elsholtzia, mint shrub
shrub   zone 5b   Aug-Sep   150 cm   N
Value for honey: claimed to "swarm" with bees[72]; *E. crispa* and *E. cristata* are also favoured by bees[35].
Notes: spikes of fragrant, lilac-purple flowers; needs full sun; not heavy clay soil; one of the few fall flowering shrubs.

Enkianthus campanulatus (Miq.) Nichols.
redvein enkianthus
shrub   zone 5   mid-May   2.5-4 m   N(P?)
Value for honey: never fails to attract honeybees; "dearly loved by bees" in Charlottetown, P.E.I.[a]. Several related ericaceous plants produce a H known to be poisonous to humans (see Section 10).
Notes: ericaceous shrub with drooping clusters of bell-shaped flowers that resemble those of the blueberry; flowers appear just before the leaves; autumn colour is scarlet; introduced to Canada.

Erica L. (see individual spp. for more detail)
heath, bruyère
shrub   all year   NP
Value for honey and pollen: majority of cvs. are relished by bees[35]; *E. tetralix* L. (crossleaf heath) is said to have a corolla too long for the honeybee to collect N (it is essentially a bumblebee's flower)[72].

---

[a] Brehaut, K. (1986). *Personal communication.*

Notes: peaty or acid soil; ericaceous, low growing plants.

Erica carnea L.
spring heath, bruyère herbacée
shrub   zone 6   Jan-May   30 cm   NP
Value for honey: see *Erica*.
Value for pollen: P available all day[17].
Honey: is light to dark yellow in colour; aroma is very sharp[17].
Notes: peaty soil; sun; rosy red or white flowers.

Erica cinerea L.
Scotch heath, twisted heath, bell heather, bruyère cendrée
shrub   zone 7   mid-Jun-early Jul   45 cm   NP
Value for honey: see *Erica*.
Honey: is brownish to port-wine-red in colour; granulates rapidly; flavour is characteristic[17].
Notes: acid soil; flowers purplish red; moors and open woods, local, Nantucket I., Mass.[23].

Euonymus alata (Thunb.) Siebold
winged spindle tree, winged euonymus, fusain ailé
shrub   zone 3   May-Jun   120 cm-2.7 m   N[36]
Notes: autumn colour and ornamental fruits.

Evodia daniellii (J. Benn.) Hemsl.
Korean evodia
tree   zone 6b   end Jul-early Aug   5-9 m   N(P?)
Value for honey: has an excellent reputation among beekeepers.
Notes: small white flowers in large clusters; weak-wooded.

Fagus sylvatica L.
beech, European beech, hêtre commun
tree   zone 6   27 m   D(P)
Value for honey: a major source of honeydew H in mid Europe[17]; may be more attractive to bees after rain[17]; considered a troublesome source of honeydew in the UK.[35].
Value for pollen: bees known to visit for P, but P is probably not of much consequence[35]; does not flower every year.
Notes: excellent ornamental shade tree; several fine cvs.; woolly beech aphid is a common pest of this sp., and the likely cause of honeydew.

Fuchsia magellanica Lam.
Magellan fuchsia, hardy fuchsia
shrub   zone 8   mid-Jun   150 cm   NP
Value for honey: considered a dependable yielder; rich in N; pendant flowers are protected from rain or dew[35].
Honey: H from fuchsia said to be light in colour with little flavour[35].
Notes: grown in only the mildest parts of B.C. as an herbaceous per.; profuse bloom.

Gaultheria shallon Pursh
salal, shallon
shrub   zone 7   Jun-mid-Jul   -3 m   N
Value for honey: regarded as one of the best native N plants in w. Canada, especially coastal B.C.[72,73].
Notes: showy racemes of white flowers; suitable for foundation planting and shrub borders; s. Alaska, B.C. to s.Calif.[23]; see Table 18.

Gaylussacia brachycera (Michx.) A. Gray
box huckleberry
shrub   zone 5   Jun   30 cm   NP
Value for honey: a very attractive source of N; *G. baccata* (black huckleberry) is seldom cultivated, but is a recognized H plant in B.C.[72,73].
Notes: evergreen and slow growing ericaceous shrubs; foliage and fruit have ornamental value; sandy woods and slopes, E. to se. USA.[23].

Gelsemium sempervirens (L.) Ait.
Carolina jasmine, evening trumpetflower
vine   -15° to -12° C   early Mar-May   (N)P
Value for honey and pollen: visited by foraging bees; accounts of animals poisoned by eating this plant, a single account of H (containing P) causing toxicity in a few people, and reports of extensive fatalities of young bees in colonies foraging on this sp.[7,72]; see Section 10.
Notes: handsome per. vine, fragrant yellow flowers with long funnel shaped corollas; borderline hardy in the mildest parts of B.C.; native in parts of s. USA.[23].

Gleditsia triacanthos L.
common honey locust, thorny locust, honeyshuck, fevier d'Amerique, carouge à miel
tree   zone 4   mid Jun   15-20 m   N
Value for honey: HP3[16]; N flow is too short for large yields, so surplus is usually unknown; bloom lasts about 10 days[17]; has been recommended for roadside planting to increase bee forage (i.e. N.J.)[62].
Notes: spreads and can become a nuisance; useful for hedges and windbreaks or erosion control[17]; rich woods, w.N.Y. and S., also cultivated and becoming established to N.S. and N.E.[23].

Hamamelis mollis D. Oliver
Chinese witch hazel, hamamélide de Chine
shrub   zone 6b   Mar-Apr   3-7.5 m   P[36]
Notes: unexcelled for the effect of its fragrant yellow flowers.

Hamamelis virginiana L.
witch hazel, café du diable
shrub   zone 4b   early Oct   3-4 m   P[35]
Notes: a large shrub, extremely attractive in flower and is virtually the last shrub to bloom in e. Canada; dry to moist woods, s. Que., N.S., N.E.[23]; astringent bark and leaves reputedly have medicinal properties[51].

TREES AND SHRUBS   83

Hebe Comm. ex Juss.
hebe
shrub   N
Value for honey: several spp. cultivated in Europe are observed to be attractive for N[35]; not commonly grown in Canada.

Hydrangea L.
shrub   Jul-Sep   NP[36]
Value for honey and pollen: generally observed to be attractive to honeybees for N and P[72]; no specific references to differences in attractiveness between spp. recorded.

Hypericum L.
St. John's wort, millepertuis
shrub, ground cover   Jul-Aug   P[36]
Value for pollen: attractive for P[35].
Notes: some of these make excellent ground covers; bright yellow summer flowers.

Ilex aquifolium L.
English holly, European holly, Oregon holly, houx
shrub   zone 7   Jun(?)   3-9 m   N(P, male plants only)
Value for honey: both male and female flowers (on separate plants) secrete N freely for 2-3 weeks[35]; usually blooms between fruit and clover[35]; small inconspicuous flowers said to "swarm" with bees, but unfortunately they do not flower freely[35].
Notes: grown commercially in orchards in parts of B.C., and N.S. for Christmas decorations.

Ilex glabra (L.) A. Gray
inkberry, gallberry, Appalachian tea, winterberry
shrub   zone 5b   Aug (N.S.)   1.2-5.4 m   N(P male plants only)
Value for honey: H yield is usually 14-23 kg/colony/season and in an exceptional year, is up to 135 kg/colony/season[17]; said to be difficult to overstock inkberry shrubberies with bees (apiaries of 100-300 colonies placed on inkberry stored a surplus)[72]; N flow best in dry weather[17]; both male and female plants yield N.
Honey: is classed as white to extra light amber; very slow to granulate; heavy body; desirable aromatic after taste[17].
Notes: low sandy or peaty soils, N. to Mass., Isle au Haut, Me., N.S.[23]; greater commercial use could be made of this native holly[81].

Ilex opaca Ait.
American holly, houx
shrub   zone 7   all May   1.8-7.5 m   N(P male plants only)
Value for honey: a very attractive N source; bees able to gather up to 67 kg/colony in the 10-12 days of bloom (s.USA.)[72]; both male and female plants yield N.
Honey: said to have a pleasant "twang"[72].
Notes: not hardy at Ottawa; of easy culture in parts of B.C., milder regions of the Maritimes, and s.Ont., winter protection is recommended, especially in the Maritimes[81].

Ilex verticillata (L.) A. Gray
winterberry, black alder, aune blanche, houx verticillé
shrub   zone 3b   Jun(?)   1.5-2.4 m   N(P male plants only)
Value for honey: noted as an important N plant; both male and female plants yield N[72].
Notes: deciduous; the most widespread of the American hollies; cvs. found in Ont., N.S. and Nfld.[23].

Kalmia angustifolia L.
sheep laurel, lambkill, pig laurel, wicky, dwarf laurel, kalmie à feuilles étroites
shrub   zone 1   mid Jun(?)   90 cm   N
Value for honey: worked well for N and is an important source in N.S.; *K. latifolia* (mountain laurel) contains a toxin, andromedotoxin, which accumulates in H from this plant, and is poisonous to humans[7]; see Section 10.
Notes: pink flowers in clusters in leaf axils; acid soil; disliked by farmers because it is poisonous to animals, especially to sheep; not generally cultivated; found in dry or wet sites; sterile soil, old pastures and barrens, Lab. to Man., S. to Nfld., N.S., N.E. and S.[23].

Kalmia latifolia L.
mountain laurel, calico bush, ivybush, spoonwood, kalmie à larges feuilles
shrub   zone 5b   mid Jul (Mar-May in B.C.?)   N
Value for honey: not heavily worked; a source of poisonous H[72]; a toxin, andromedotoxin, can be detected in the H and is poisonous to humans; not clear whether or not this plant is safe for bees[7]; see Section 10.
Notes: beautiful flowering evergreen shrub, native to mountainous areas of e. N. America where soil is acid; rocky or gravelly woods and clearings, sometimes swamps, in acid or sterile areas, N. to N.E.[23]; also a very popular garden shrub.

Koelreuteria paniculata Laxm.
golden rain tree, varnish tree, koelreuterie paniculée
tree   zone 6b   mid Jul   5-9 m   N
Value for honey: heavily worked for N[50].
Notes: loose bunches of yellow flowers; grows in a wide range of soils; sometimes cultivated as a street tree.

Larix decidua Mill.
larch, European larch, mélèze
tree   zone 3b   15-24 m   D
Value for honey: an important source of honeydew H in some countries (i.e. in temperate Europe)[17]; honeydew may crystallize on tree[17]; *Cinara larcia* is a common plant-sucking pest of larch that produces a honeydew (and H) high in melizitose (i.e. 42%), causing it to crystallize very rapidly[18] (melizitose, a trisaccharide, comes from the French word for larch, mélèze[18]); melizitose is one of the less soluble sugars in H, and bees collect only when it is liquid, in the morning or when humidity is high[18].
     For other characteristics of melizitose of interest to beekeepers, see mention under *Pseudotsuga menziesii* (Douglas fir) and in Section 9.
Honey: granulation is reported as rapid, and also as slow[17]; fairly sweet[17]; see Section 9.
Notes: one of the most popular of larches for ornamental use; native to temperate

Europe and Asia; yellow in autumn, with deciduous needles; well drained, light gravelly soil; low-lying areas are not tolerated by most larches; *L. larcina* (Du Roi) C. Koch (eastern larch, épinette rouge) is native to Canada, but is value as a H source has not been well documented.

Laurus nobilis L.
laurel, bay, sweet bay, laurier sauce
shrub   zone 9a   Jun   3.6-4.5 m   N
Value for honey: flowers are rich in N[3].
Notes: often grown in tubs and brought indoors before the first frosts; aromatic leaves are used in cooking and also processed to yield an essential oil.

Lavandula angustifolia Mill.[51]
(syn. *Lavandula officinalis* Chaix in Vill.; *L. spica* L.; *L. vera* DC; *L. pyrenaica* DC.)[51]
lavender, true lavender, lavande commun, lavande véritable, aspic, spic
shrub   zone 7   Jul-Aug   less than 1m   N(P)
Value for honey: HP3[17]; H yield 5-20 kg/colony/season[17]; N sugar concentration 14-33%[17]; an excellent N plant where hardy.
Honey: classed as extra light amber to dark amber; is yellow or golden in colour; granulation is very smooth, like butter; good body; flavour is highly regarded and delicate[17].
Notes: needs chalky well-drained soil; full sun; a highly desirable fragrant shrub for hedge or rock garden.

Lavandula × intermedia Emeric ex Loisel.[51]
(syn. *L. × aurigerana* Mailho.; *L. × burnati* Briq.; *L. × hybrida* Revercheron ex Briq.; *L. × hortensis* Hy; *L. × feraudi* Hy; *L. × guilloni* Hy; *L. × leptostachya* Pau; *L. × senneni* Fouc. ex Chaytor)[51]
lavandin, lavande bâtarde
shrub   temperate   N(P)
Value for honey: HP4[17]; N sugar concentration 38-67%[17]; bees induce a noticeable increase in yield of lavender oil through their foraging[54].
Honey: classed as white; granulation is very fine; aroma is strong; a specialty H that is in demand (i.e. lavandin H)[17].
Notes: cultivated commercially for oil used in perfumery; dry, sunny site[17].

Ledum groenlandicum Oedr.
Labrador tea
shrub   zone 1   late May-early Jun   90 cm   N
Value for honey: an important source of H in some provinces, especially Man., n.Ont. and B.C.[72]; *L. glandulosum* Nutt. is reported to be an important source in B.C.
Honey: H has a minty flavour; *L. palustre* L. (wild rosemary, crystal tea, lédon des marais) produces a N, which when made into H is poisonous to humans (the toxin is glucoside ericolin)[7]; see Section 10.
Notes: clusters of white flowers; evergreen leaves; not commonly cultivated; peaty soils, Greenl. and Lab. to Alaska S., especially in bogs to Nfld., N.S., N.E.[23].

TREES AND SHRUBS

Lespedeza bicolor Turcz.
shrub   zone 4   late Jul-Aug   60-90 cm   N
Value for honey: one of the most attractive of the lespedezas for bees[72]; N sugar concentration 49%[17]; may take a few years to establish a new plant before it will begin blooming[72]; not considered dependably attractive to bees[72].
Notes: luxuriant bloom consists of small clusters of rosy-purple flowers; will grow in most soils, including those slightly acid (unusual for a legume); suitable for rough ground, erosion control or for ornamental planting in herbaceous border; usually killed to the ground by frost, but resprouts in spring.

Lespedeza cryobotria Miq.
shrub   2-3 m   N
Value for honey: more consistently attractive to bees than either *L. bicolor* (shrub bush clover) or *L. thunbergii* (Thunberg lespedeza)[72]; H yield 26 kg/colony/season[17].

Lespedeza thunbergii (DC.) Nakai
Thunberg lespedeza
shrub   zone 5   late Sep   4.8 m   N
Value for honey: as attractive as *L. bicolor* (shrub bush clover)[72].
Notes: reddish-purple flowers; spreading in cultivation, Mass. and S.[23].

Leucothoe D. Don
leucothoe, fetter bush
shrub   Jun   90-180 cm   N(P)
Value for honey: valuable in some areas, but H is suspected of being toxic to humans[50] (see Section 10).
Honey: several related ericaceous plants produce H known to be poisonous to humans, e.g. some *Rhododendron*, and *Kalmia latifolia* (see Section 10).
Notes: deciduous and evergreen shrubs belonging to the heath or ericaceous family; *L. fontanesiana* (Stued.) Sleum. (drooping leucothoe) is the most ornamental of the group, and there are several garden cvs.; require protection in N., thrive in most soil of peat and sand.

Ligustrum japonicum Thunb.
Japanese privet, troène du Japon
shrub   zone 9a   Jun(?)   3 m   NP
Value for honey: where this is widely spread (e.g. Texas), a 2 1/2 super surplus of acceptable H reported; privet H has a very poor reputation[35] for palatibility (see Section 8).
Notes: is often considered the most handsome of the deciduous privets when in flower; graceful habit and panicles of white flowers.

Ligustrum vulgare L.
common privet, troèn, raisin des chien, fresillon
shrub   zone 5b   Jun-Jul   3-4.5 m   NP
Value for honey: yields N freely[35].
Honey: dark in colour; thick; bitter and liable to spoil other H if mixed with the main crop to any extent[35] (usually not enough collected to have a detrimental influence[35]); see Section 8.
Notes: found naturally from s.Me., s.Ont., and N.E. in thickets and open woods[23].

Liriodendron tulipifera L.
tulip tree, tulip poplar, yellow poplar, whitewood, tulipier de Virginie
tree zone 5b mid Jun 12-24 m N(D)[17]
Value for honey: H yields of 35-45 kg/colony/season reported[17]; N may "rain" from trees[17]; usually more N secreted than can be collected by honeybees and other insects[35]; initial concentration of N is low, 17%, but higher on the second day of flowering (i.e. 36%)[17]; each flower secretes for only about 2 days[35], and N may only be secreted in warm weather; trees do not begin flowering until they are about 15 years old[17].
Honey: reddish, but becomes less so during storage; classed as light amber to amber; heavy body; flavour distinctive[17].
Notes: long-lived desirable ornamental tree with large white tulip-like flowers; rich soil, Mass., to s.Ont., Wisc., and S.[23].

Lonicera L. (see individual spp. for more detail)
honeysuckle, chèvrefeuille
vine, shrub summer NP
Value for honey: most *Lonicera* have corollas so deep that N is accessible only to humming-birds and certain moths[35]; spp. listed here have shorter corollas and thus are more useful as bee forage. *L.* × *purpusii* Rehd. (winter flowering honeysuckle) and *L. standishii* Jacques. are also attractive to honeybees, but are not listed[35].
Notes: a sturdy group of shrubs and vines, with very few disease problems or pests; generally all honeysuckles thrive in full sun; ornamental berries are eaten by birds.

Lonicera involucrata (Richardson) Banks ex K.Spreng.
bearberry honeysuckle, inkberry, twinberry
shrub zone 1 Apr-Jun (to Aug?) 120 m NP
Value for honey: a source of surplus in B.C., especially on the coast[72] (this claim is refuted by a B.C. beekeeper[a]); see *Lonicera*.
Notes: yellow flowers; plants are loose and straggly; suited to the wild-type garden; found naturally on banks of streams and swamps, calcareous woods, Gaspé Pen., Rimouski Co. and Laurentide Park, Que., and ne.N.B., James Bay reg. to L. Superior reg. of Ont.[23].

Lonicera morrowii A. Gray
Morrow honeysuckle
shrub zone 4 late May-Jun 1.8 m NP
Value for honey: useful for early brood rearing stimulation where sp. is naturalized[72]; see *Lonicera*.
Notes: grows only in the mildest plarts of B.C., and usually as a clipped hedge; will withstand salt spray; see *Lonicera*.

Lonicera tatarica L.
Tatarian honeysuckle, chèvrefeuille de Tartaire
shrub zone 2 late May-early Jun 2.7 m NP
Value for honey: considered a good N source; N sugar concentration 45%[94]; bees work these vigorously for 2 weeks[94]; prolific bloom.
Notes: profuse white to pink flowers; very vigorous, hardy and tolerant of most soil;

---

[a] Mitchell, S. (1987). *Personal communication.*

most commonly grown honeysuckle in Canada; thickets and borders of woods, shores, etc., escaped from cultivation in Que., and Ont., S. to N.E. and S. beyond Can. limits[23]; see *Lonicera*.

Lonicera xylosteum L.
European fly honeysuckle, chèvrefeuille des boissons, chèvrefeuille à balais
shrub   zone 2   Apr-Jun(?)   1.2-2.4 m   NP
Value for honey: corolla is only 3 mm long, so all the N is available to honeybees[35]; see *Lonicera*.
Notes: yellowish white flowers, followed by bright showy berries; roadsides and thickets, escaped from cultivation, N.E. to Mich., and S.[23]; see *Lonicera*.

Magnolia acuminata (L.) L.
cucumber tree, arbre aux concombres
tree   zone 5   May-Jun   27 m   (N)DP
Value for honey: produces a heavy honeydew flow in May and June in parts of e. Canada and s.Ont[72].
Notes: fast growing; inconspicuous greenish-white flowers; rich woods, w.N.Y. and s.Ont. and beyond Can. range[23].

Magnolia grandiflora L.
southern magnolia, magnolia à grandes fleurs
tree   zone 9a   early Jun-Sep   4.5-6 m   (N,D?)P
Value for honey: surplus H sometimes reported, but not considered particularly attractive to honeybees[72].
Honey: said to produce a dark, strong H in some areas, may be unpalatable[17] (see Section 8).
Notes: grows only in the most sheltered locations of coastal B.C.; fragrant white flowers.

Magnolia × *soulangiana* Soul.-Bod. (: *M. heptapeta* × *M. quinquepeta*)
saucer magnolia, Chinese magnolia, magnolia de Soulange
tree   zone 5b   May-Jun   4.5-7.5 m   P[36]
Value for pollen: not conspicuously attractive, and for P only[36].
Notes: flower buds can withstand temperatures as low as -25° C.

Magnolia stellata (Siebold & Zucc.) Maxim.
star magnolia, magnolia étoilé
tree   zone 7b   late Apr   6 m   P[36]
Value for pollen: not conspicuously attractive, and for P only.

Mahonia aquifolium (Pursh) Nutt.
(syn. *Berberis aquifolium* Pursh)
Oregon grape, mountain grape, holly mahonia, holly barberry, mahonia à feuilles de houx
shrub   zone 5   May   90 cm-1.8 m   (N)P
Value for pollen: considered useful for early brood rearing[50].
Value for honey: worked well for N[35].
Notes: planted as an ornamental for its yellow flowers, berries and broad leaved evergreen foliage; sometimes incorrectly called a *Berberis*; see Table 18.

Malus Mill. (see individual spp. for more detail)
apple, pommier
tree   zone 2b-5   late Apr-late May   N(D)P
Value for honey: HP1/2[16] (for *Malus, Pyrus* and *Prunus*); beekeepers generally agree that apple is a superior source of N to *Pyrus* (pear) or *Prunus* (plum, cherry)[35,72]. One important reason for this good reputation is that *Malus* flowers slightly later than *Pyrus* or *Prunus*, when colonies are likely to be strong enough to store a surplus. Strong colonies in B.C. have been reported to store as much as 45 kg/colony/season, and gains of 1.3-3.6 kg/colony/day are not uncommon[50].

An apple tree's bloom may last for about 14 days (average 9 days), but there is a great deal of difference in flowering times of cvs. The total blooming period for *Malus* is about 4 weeks[35,54]. The later flowering spp. are often useful in bridging the June gap in the N flow[35].

Apple flowers like most flowers in the *Rosaceae* (rose) family, have a very open shape. As a result of this, the N is liable to be diluted by rain or dew[35]. In fine weather, and as the day progresses, evaporation tends to increase the N sugar concentration, making the flowers increasingly attractive to bees.
Value for pollen: attractive for P which is collected in the mornings only (like most *Rosaceae*)[83].
Honey: (for *Malus, Prunus, Pyrus*) is light in colour; granulation is said to be quick; soft, fine grains; flavour is excellent, delicate; aroma is fine[16].
Notes: The trees of this genus constitute the apples and crab apples. Like many woody plants, some spp. and cvs. in this group have the disconcerting habit of bearing large crops of flowers and fruits only in alternate years. As there is little that can be done to change this, only varieties that bear annually should be chosen if this is important to the grower.

Some spp. and cvs. are less susceptible than others to apple scab and fire blight diseases[95] and they require less care and fewer pesticides.

Malus baccata (L.) Borkh.
Siberian crab apple, Siberian crab, pommier à baies
tree   zone 2b   early May   5-9 m   NP[17]
Value for honey: see *Malus*.
Honey: see *Malus*.
Notes: profuse fragrant flowers; cultivated as an ornamental.

Malus coronaria (L.) Mill.
wild crab, American crab apple, sweet-scented crab, wild sweet crab, pommier odorant
Value for honey and pollen: see *Malus*.
Notes: native to Canada; ideal bee pasturage where it is common; note that several of the cvs, available are double and should not be grown for bees; found in bottomlands, wooded slopes, thickets and clearings, centr.N.Y. and s.Ont. to s.Wisc. and beyond Can. range[23].

Malus domestica Borkh.[17]
(syn. *Malus communis* Poir. in part; *Malus sylvestris* (L.) Miller var. *domestica* (Borkh.); *Pyrus malus* L. in part)[17]
apple, pommier
tree   -37° to -29° C   13.5 m   N(D)P
Value for honey: HP1/2[16]; H yield 36 kg/colony/season[17]; N sugar concentration 30-

87% (the latter in particularly hot climates)[16]; H yield per day is 1.3-3.6 kg/colony[17], but H is stored only when foraging conditions are favourable[17]; see *Malus*.
Notes: botanists disagree as to what constitutes this sp. however it is rarely cultivated in its pure form[17]. It is included here because its value as a bee plant has been better researched than other spp. of *Malus*.

Malus pumila Mill.
(syn. *Malus communis* Poir.; *M. domestica* Borkh.; *Pyrus pumila* (Mill.) C. Koch, not J. Neumann ex Tausch)
common apple, pommier commun
tree   -37° to -29° C   early May   13.5 m   NP
Value for honey: N sugar concentration 50%[94]; see *Malus*.
Value for pollen: see *Malus*.
Notes: considered to be a parent of the modern apple cvs.[95]; resembles *M. sylvestris*, but is less thorny.

Myrica gale L.
sweet gale, bog myrtle, meadow fern, piment royal, bois sent bon, gale odorant
shrub   zone 2   Mar-Apr   120 cm   P[20]
Notes: flowers are catkins; not often used ornamentally; moist peaty soil best; found in shallow water and swamps. Lab. to Alaska, S. to Nfld., N.S., L.I., s.Ont., Ore. and beyond Can. range[23].

Oxydendrum arboreum (L.) DC.
sourwood, sorrel tree, andromède en arbre
tree   zone 7   Jul   6-12 m (taller when cultivated)   N
Value for honey: H yield up to 102 kg/colony/season (mean yield of 271g)[17]; heaviest yield very 4-5 years[17], but generally considered to be a dependable source[30].
Honey: is yellowish in colour, with a pinkish cast; classed as almost water-white to extra light amber; slow granulation; heavy body; flavour delicate, slightly piquant or sour[17].
Notes: a superior ornamental tree; blooms for 2-3 weeks[17]; scarlet autumn colour.

Paulownia tomentosa (Thunb.) Steud.
paulounia, princess tree, karri tree, paulownie impériale
tree   zone 7   mid-May   5-10 m   N(P?)
Value for honey: reported to be highly attractive to bees[72].
Notes: profuse bloom; flower buds may winter-kill near the extremes of its range[95]; introduced and naturalized N. to s.N.Y.[23]; if grown beyond its hardiness range will die back to ground each winter and resprout the next spring but will not flower.

Perovskia atriplicifolia Benth.
silver sage
shrub   zone 5   Aug-Sep   90-150 cm   N
Value for honey: said to be covered with bees collecting N, especially in warm weather[35].
Notes: masses of deep lavender flowers, dies to the ground each winter, and so should be treated as a herbaceous perennial; belongs to the mint family (*Labiatae*).

Phellodendron amurense Rupr.
Amur cork tree, arbre à liège de l'Amur
tree   zone 3   May-Jun   15 M   N(P from male trees only)
Value for honey: HP4[17]; N secretion is maximum at temperatures between 21° and 23° C, and RH 58%[17]; one report states that H from its honeydew, or the honeydew itself, is lethally toxic to bees[7] (see Section 10).
Notes: sexes are separate; a useful open shade tree.

Picea abies (L.) Karst.
(syn. *Picea excelsa* (Lam.) Link)
Norway spruce, épinette Norge
tree   zone 2b   18-30 m   D
Value for honey: source of honeydew H; HP 20-500 kg/ha (insect not specified)[17]; mean H yield 40 kg/colony/year[17]; H noted to be a safe winter food for bees[17]. Bees observed working ornamental specimens in Guelph, Ont. in large numbers. This was noted during a season in which the flow from this came before clover and H was thought to be of acceptable quality for table use (insect was spruce bud scale, *Physokermes picea*)[72].
          In West Germany, *P. picea* (insect) produced a heavy flow with up to 40 kg/day/colony, with a water content of 14%[17], but no quantitative information for Canada available. See Section 9.
Honey: dark reddish brown to greenish black in colour; granulation slow (from 2 different insects); flavour fairly sweet[17]; see Section 9.
Notes: widely-planted to over-planted ornamental specimen; fast growing when young, does not grow old gracefully; many cvs.

Pieris floribunda (Pursh ex Sims) Benth. & Hook.
pieris, mountain andromeda, fetterbush
shrub   zone 5   late Apr-May   180 cm   N(P?)
Value for honey: bees do visit these flowers, but as the corolla is long, and its opening constricted, honeybees cannot obtain much N unless the secretion is very heavy[35]; honeybees also visit *P. japonica* (lily of the valley bush) in the lower mainland area of B.C.[a].
Honey: H suspected of being poisonous to humans[34] (see Section 10).
Notes: an evergreen ericaceous shrub that will grow in acid or alkaline soil; flowers drooping in upright terminal spikes; found naturally on moist hillsides beyond Can. range[23], but is grown ornamentally in Canada.

Pinus sylvestris L.
Scots pine, Scotch fir, Scotch pine, pin
tree   zone 2   12-27 m   D
Value for honey: flow may be very heavy, and in some areas, bees are moved to forests for this flow (i.e. mid Europe)[17]; honeydew potential on *Pinus* is 10 kg/ha[17]; 60% of Greek H is from the insect, *Marchalina hellenica* on pines (i.e. partly *P. sylvestris*)[17]; see Section 9.
Notes: attractive coniferous tree; typical specimen has twisted bluish green needles in bundles; picturesque, open habit; much cultivated and locally naturalized from N.E. to Ont., S. to beyond Can. range[23].

---

[a] McCutcheon, D. (1986). *Personal communication.*

Populus L.

poplar, aspen, cottonwood, peuplier

tree   Mar-Apr   (D)P

Value for honey: honeydew may be collected[72]; crystallized honeydew found to contain 40% melizitose (in Austria, insect not specified)[17]   see entry under *Pseudotsuga menziesii*, Douglas fir, and also see Section 9 for more information on melizitose, a very insoluable trisaccharide of importance to beekeepers.

Value for pollen: sometimes considered of value for its early P; P is poor in protein[83].

Notes: several spp. native to Canada, *P. temuloides* (trembling aspen) being the most common; *Populus* spp. are typically fast growing and weak-wooded; usually not desirable in urban areas as roots clog drains and suckers present a nuisance.

Prunus L. (see individual spp. for more detail)

cherries, plums, almonds, peaches, apricots, prunier, cerisier

trees and shrubs   zones 1-7b   late Apr-mid May   NP

Value for honey: HP1/2[16] (for *Malus*, *Pyrus*, *Prunus*); all are generally valued as early sources for spring build up; H surplus rarely obtained because most flower too early, when colony populations are low and weather conditions may be unfavourable for foraging.

The many ornamental, double flowered cvs. are not much visited by bees. See Table 18.

Value for pollen: an excellent protein source[83]; all commercially grown *Prunus* are largely dependent on honeybees for pollination and economic fruit set[54].

Honey: (for *Malus*, *Pyrus*, *Prunus*) is light in colour; granulation is quick; grain is fine; flavour is excellent, delicate[16].

Notes: ornamental and commercial *Prunus* are particularly susceptible to various insect and disease pests, and thus often have to be treated with pesticides. Beekeepers wishing to grow *Prunus* primarily for bee forage should choose those spp. requiring less attention and fewer sprays. Refer to Section 11 for more information on pesticides and their relative hazards to honeybees.

The following list of *Prunus* in order of bloom includes species for which there are no recorded nectar or honey values. Nevertheless, these plants are likely to be attractive to honeybees. For such species approximate range or habitat and naming authority are given in the list below.

VERY EARLY: *P. armeniaca* (flowering apricot, abricotier); *P. cerasifera* (Myrobalan plum); *P. domestica* (garden plum, prunier cultivé); *P. incisa* (Fuji cherry); *P. sargentii* (Sargent cherry); *P. subhirtella* (Higna cherry); *P. yedoensis* (Yoshino cherry); *P. besseyi* L.H. Bailey[23] (sand cherry), sand hills, open plains, rocky slopes, Man. to Wyo. and beyond[23]; *P. mahaleb* (L.) (perfumed cherry, cerisier odorant), roadsides and borders of woods, rocky banks etc., N.E. to s. Ont. and S.[23].

EARLY: *P. americana* Marsh.[23] (American plum, wild plum), thickets and borders of woods, stream banks and fence rows, n. to w. N.E., N.Y., s. Ont., s. Man., Mich., Wisc. and beyond Can. limits (blooms early June in n.)[23]; *P. avium* (Mazzard cherry, cerisier de France); *P. padus* (European bird cherry); *P. pensylvanica* L. (pincherry, wild red cherry, cerisier d'été), dry woods, recent burns and clearings s. Lab. Pen. to B.C., S. to Nfld., N.S., N.E., L.I. and beyond Can. range (blooms early July in n. areas)[23]; *P. persica* (flowering peach, pêcher); *P. serrulata* (Japanese cherry).

MODERATELY EARLY: *P. cerasus* (sour cherry, cerisier à grappes); *P. depressa* Pursh[23] (sand cherry, gogoune), gravelly or sandy soil or beaches in calcareous soil, Gaspé Pen. to L. Mistassini, Que., W. to Manitoulin Distr., Ont., S. to rivers of N.B.,

n. N.E. and beyond Can. limits[23]; *P. virginiana* (western common choke cherry, cerisier).

LATE: *P. serotina* Ehrh.[23] (black cherry, cerise d'automne), dry woods and fence rows, n. N.S., N.B., s. Que., s. Ont., etc. (blooms in June in N.)[23]; *P. susquehanae* Willd.[23] (sand cherry), sand or other acid, dry to wet open habitats, sw. Me. and sw. Que. to se. Man., S. to L.I. and beyond Can. range[23].

VERY LATE: *P. nigra* Ait.[23] (Canada plum, guignier), thickets, stream banks, borders of woods, etc., Que. to Man., S. to N.S. (introd.), N.E., and beyond Can. range[23]; *P. pumila* L.[23] (sandy cherry), dunes and sands or calcareous rocky shores, Ont., St. Lawrence basin, N.Y. and S.[23].

Prunus armeniaca L.
apricot, abricotier
tree   zone 7   late Apr   7.5 m   NP
Value for honey: HP2[16]; see *Prunus*.
Value for pollen: see *Prunus*.
Notes: see *Prunus*.

Prunus avium (L.) L.
sweet cherry, Mazzard cherry, cerisier de France, guignier sauvage
tree   zone 4b   early May   180 cm   NP
Value for honey: HP2[16]; N sugar concentration 28% (average)[54]; considered to be second only to *Malus* (apple) among N producing tree-fruits[5]; see *Prunus*.
Value for pollen: see *Prunus*.
Honey: see *Prunus*.
Notes: found in roadside thickets and borders of woods, P.E.I. to Mich. and beyond Can. range[23]; introduced and naturalized[23]; see *Prunus*.

Prunus domestica L.
garden plum, European plum, common plum, prunier cultivé
tree   -29° to -23° C   late Apr   9m   NP
Value for honey: HP2[16]; some observers believe that honeybees usually prefer plums to *Pyrus* (pear) or *Ribes* (currant)[35]; others hold that plums are not very attractive, but consider them to be significant sources for brood rearing stimulation[54]; N sugar concentration increases during the day, or as RH decreases (i.e. 26% at RH 100% at 0700-0800h, and 62% at 1400 with RH 53%)[54]; honeybees working plum shifted from plum to *Arctostaphylos* (manzanita) at 1000h, but returned to plum after 1200h[54].
Value for pollen: see *Prunus*.
Honey: see *Prunus*.
Notes: this is the common "old-fashioned" plum of horticulture; there is 3-4 weeks between the earliest and the latest flowering plum cvs.; see *Prunus*.

Prunus dulcis (Mill.) D.A. Webb
(syn. *Amydalis communis* L.)
flowering almond, amandier
tree   zone 7   late Apr   6 m   NP
Value for honey: HP1[16]; eagerly worked for N all day[54]; see *Prunus*.
Value for pollen: P is collected all day, but foraging is most active around noon[54]: see *Prunus*.
Notes: flowers are pink and white, 3-5 cm in diameter; see *Prunus*.

Honey: bitter and considered of poor quality; used in the bakery trade[72]; see *Prunus*.

Prunus laurocerasus L.
cherry laurel, laurier cerise
tree   zone 7   late May   1.8-9 m   NP
Value for honey: also produces extra-floral N (from nectaries on under-surfaces of leaves) and is visited all summer, depending on competition from other forage[35].
Value for pollen: see *Prunus*.
Notes: introduced and not reliably hardy in s.Ont.; evergreen; see *Prunus*.

Prunus padus L.
European bird cherry, bird cherry, hagberry, cerisier à grappes, bois puant
tree   zone 2   early May   3-12 m   N[35]
Value for honey and pollen: see *Prunus*.
Notes: large fragrant flowers in drooping racemes; introduced and spread from cultivation to roadside thickets.

Prunus persica (L.) Batsch.
(syn. *Persica vulgaris* L.)
peach, pêcher
tree   zone 6b   early May   6 m   NP
Value for honey: average N sugar concentration 39% in floral N[79]; extra-floral N sugar concentration 48%[94] (extra-floral nectaries are found in the axillary region where the petiole attaches to the stem); extra-floral N may be secreted before, after, and during blooming; see *Prunus*.
Value for pollen: see *Prunus*.
Honey: see *Prunus*.
Notes: widely cultivated in many countries for fruit, and in Canada, in B.C. and Ont. in particular; see *Prunus*.

Prunus sargentii Rehd.
Sargent cherry
tree   zone 5   late Apr   9-14 m   NP[36]
Value for honey and pollen: see *Prunus*.
Honey: see *Prunus*
Notes: very intolerant of air pollution; see *Prunus*.

Prunus serrulata Lindl.
Oriental cherry, Japanese flowering cherry, cerisier à feuilles dentées en scie
tree   zone 6   mid to late May   5-8 m   NP[36]
Value for honey and pollen: singles and semi-doubles only are visited by bees[35]; see *Prunus*.
Honey: see *Prunus*.
Notes: see *Prunus*.

Prunus spinosa L.
blackthorn, sloe, épine noire, prunellier
tree   -29° to -23° C   late May   3.6 m   NP
Value for honey: HP1[16]; see *Prunus*.
Value for pollen: see *Prunus*.

Honey: see *Prunus*.
Notes: introduced fruits used for flavouring liqueurs; see *Prunus*.

Prunus virginiana L.
western common choke cherry, cerisier de Virginie, cerisier noir
tree   zone 2   mid May   4-8 m   NP
Value for honey and pollen: seldom provides surplus[94]; see *Prunus*.
Honey: is light in colour and mild[94]; see *Prunus*.
Notes: thickets, shores and borders of woods, Nfld. to Sask., S. to N.S., N.E. and beyond Can. range[23]; see *Prunus*.

Prunus yedoensis Matsum.
Yoshino cherry
tree   zone 7b   late Apr   6-12 m   NP
Value for honey: N sugar concentration 28%[17]; see *Prunus*.
Value for pollen: see *Prunus*.
Honey: golden or pale yellow in colour; rapid granulation[17]; see *Prunus*.
Notes: most widely cultivated ornamental cherry in Japan and very popular in N. America; not known in the wild; profuse blooming and short-lived tree.

Pseudotsuga menziesii (Mirb.) Franco var. *glauca* (Beissn.) Franco
(syn. *Pseudotsuga glauca* (Mayr.) Mayr.)
blue Douglas fir (from Interior B.C.), Douglas bleu
tree   zone 3   18-45 m (up to 90 m)   D
Value for honey: honeydew H yield 22 kg/colony/season reported in Olympica Mts. (Greece), but heavy winter loss due to "dysentry" followed[72]; crystallized sugar ("fir sugar") accumulates on needles, and may fall to the ground beneath the trees[50]; important source only in hot dry weather with adequate soil moisture, full sun, and e. or n. exposure[72]; strongly claimed not to be honeydew, but rather a sap exudate directly from the tree [72] (is most probably entirely a honeydew secretion[a]); a noted source in parts of the Fraser, Thompson and Okanagan Valleys of B.C.[72]; see Section 9.
Honey: crystallizes readily and is usually considered unsafe for winter stores, but is desirable as table H[50,72]; high content (75-83%) of melizitose, a trisaccharide sugar[29]; suggested that some of the lachnids may be specially adapted to produce melizitose, and this would explain its presence in large quantities only in certain H[29].

Melizitose is one of the least soluble sugars and bees are only able to collect it while it is still liquid (i.e. in the mornings or when the humidity is high)[29]. H often granulates in the comb and may be removed and discarded by bees[29]. Honeybees are able to digest melizitose in solution by means of the enzymes in their mid hind-guts and salivary glands, but production of the necessary enzyme claimed to occur only in well nourished "summer" bees[29]. Accumulation of melizitose crystals in honeybees may damage the gut wall, causing death, and could be the reason for heavy winter losses if used for winter stores[29]; see Section 9.
Notes: excellent ornamental evergreen; slow growing; shallow rooted; needles bluish green; leaves and cones smaller than the Rocky Mts. form (*Pseudotsuga menziesii* (Mirb.) Franco).

---

[a]   Author's opinion and personal communication, R. Day (IBRA), 1986.

Ptelea trifoliata L.
shrubby trefoil, hop tree, common hop, stinking ash, wafer ash, orme à trois feuilles,
trèfle de Virginie
tree or shrub   zone 3b   May-Jun   4-8 m   N
Value for honey: very attractive for N; whole tree observed to "hum" with bees at
flowering time[35]; main N flow lasts 1 week in hot weather[35]; native to parts of s.
Ont.[72]; usually not common enough for surplus H production[50].
Notes: fruits can be used as a substitute for hops; aromatic leaves and bark; alluvial
thickets, rocky slopes and gravels sw. Que. and N.Y. to s. Ont. and beyond Can
range[23]; cultivated and spreading in N.E.[23].

Pyracantha coccinea M.J. Roem.
scarlet firethorn, petit corail, boisson ardent, pyracanthe
shrub   zone 6   May   180 cm   NP
Value for honey: N is secreted freely at times, but not during hot, dry periods[35].
Notes: usually best grown against a wall in full sun; ornamental white blossoms
followed by berries.

Pyrus L.
pear, poirier
tree   zone 2b-5b   early May   N(P)
Value for honey: HP1 [16]; honeybees have little interest in pear H and even where pear
is grown as a crop there are few or no reports of pear H[35]. N sugar concentration
considerably lower (varies from 4-25%) than that of *Malus* (apple) 42.6%, *Persica*
(peach) 28.9%, *Prunus* (plum) 25.8%, *Prunus* (sour cherry) 23.5%[54]; N concentration
depends on weather conditions[54]; flowers about the same time as apple (*Malus*) or
slightly earlier, but for a shorter length of time (i.e. flowering on a tree lasts about 1
week)[54]; see *Prunus*.
Value for pollen: produces abundant P, highly attractive to honeybees[54], and high in
protein content[83]; the number of colonies required for optimum cross-pollination has not
been determined, but 5 colonies/ha has been recommended in B.C.[11]; estimated that a
6% set of a moderately blooming tree will give a heavy crop of fruit[54]. The presence of
honeybees has been shown to be an economic advantage by increasing the number of
seeded fruit[54], even for some cvs. (e.g. "Bartlett") which produce fruit
parthenocarpically.
Honey: see *Prunus*.
Notes: all pears require persistent care due to susceptibility to numerous pests; most of
the commercial pears grown in N. America are derived from *P. communis* L. (common
pear).

Pyrus communis L.
pear, common pear, poirier commun
tree   zone 5   early May   12 m   (N)P
Value for honey: HP1 [16]; see *Pyrus*.
Value for pollen: see *Pyrus*.
Honey: see *Pyrus*.
Notes: most of the important or commercial pears grown in N. America belong to this
sp.; native to Europe and w. Asia, escaped from cultivation and can be found in
thickets, borders of woods and clearings in many areas[23].

Quercus L.

oak, chêne

tree   Apr-May   (D,N?)P

Value for honey: honeydew may be collected during the summer[72]; little or no N is collected[32]; H is not suitable as winter food for bees[17] (may refer to honeydew H).

Value for pollen: provides a large amount of P over a period of about 3 weeks[67].

Notes: a very important group of trees for both ornament and timber; several native to Canada, including *Q. alba* (white oak) in s. Que., s. Ont. *Q. macrocarpa* Michx.[23] (mossy cup oak) in w. N.B., s. Que., to Man.[23]; *Q. rubra* L. var. *borealis* (Michx. f.) Farw.[23] (northern red oak) in n. Que. to Algoma Distr., Ont., S. to N.S., and n. N.E. and beyond Can. range[23].

Quercus robur L.

English oak, truffle oak, chêne rouvre, chêne d'Angleterre

tree   zone 5   15-23 m   (N?,D)P

Value for honey: honeydew HP 20 kg/ha (spp. of insect not specified)[17]; see *Quercus*.

Value for pollen: see *Quercus*.

Honey: less dark than most other honeydew H; flavour is sweet, slightly sharp; aroma intense[17].

Notes: introduced and locally spread from cultivation to roadsides and borders of woods[23].

Rhamnus cathartica L.

common buckthorn, nerprun, noirprun, épine de cerf

shrub   -50° to -37° C   May   5.4 m   N(P?)

Value for honey: yields N freely[50]; naturalized in parts of e. Canada.

Notes: small black fruits follow the flowers, but they have little ornamental interest; is a carrier of rust of oats; bark is used medicinally as a cathartic; open woods, pastures and fence rows; following early introduction for use in hedges seed was rapidly dispersed by birds and appears as if native in Que., N.S. and beyond Can. range[23].

Rhamnus frangula L.

alder buckthorn, bourgene, aune noire

shrub   -50° to -37° C   May-Aug   5.4 m   NP[36]

Value for honey: during a 10 day period of bloom in mid May, in Guelph, Ont., scale hives gained 13 kg/colony mainly from this sp.[1]; foraging continues to a leser extent until late Aug. so if colonies were stronger at this time (May), a substantial surplus could be obtained[1]; has been recommended as a potential ground cover under high tension power lines, providing food for bees and other wildlife[1]. **Notes:** one of the most ornamental of the buckthorns; small white flowers appear all summer; not a carrier of rust of oats like *R. cathartica* (common buckthorn); fence rows and thickets, in local areas, s. Que., Ont., s. N.S., and beyond Can. limits; recently and very rapidly invasive, likely to become obnoxious in some areas[23].

Rhamnus purshiana DC.

cascara, bearberry, cascara sagrada, nerprun de Pursh

tree   -12° to -15° C   May   13.5 m   N

Value for honey: a source of H in w.B.C.[72].

Honey: is amber in colour; granulates slowly; very thick to extract; aroma is delightful[72].

Notes: small flowers and purplish berries; bark is the source of cascara sagrada, an important drug; native range is N. to Wash. on the W. coast from s. USA.[23].

Rhododendron L.
rhododendron, azalea
shrub   -31° to 0° C   Apr-late Jun   60 cm-4.5 m   NP
Value for honey: are visited chiefly by bumblebees and are not attractive to honeybees except in certain seasons when there is a lack of other forage[7,33].

    Rhododendrons like certain other ericaceous plants (e.g. *Kalmia latifolia*, mountain laurel) contain a poison called andromedotoxin, which accumulates in the H and can be lethally toxic to humans[7,44] (see Section 10).

    R. ponticum L. (Pontic rhododendron) grows wild on the shores of the Black Sea, and is claimed to be the source of H which temporarily poisoned Xenophon's soldiers in 401 BC., during the "Retreat of the Ten Thousand" from Persia[71]. Typical poisoning symptoms include nausea, faintness, vomiting and dizziness and in some cases, eventual death from respiratory paralysis[50]. Species known to have poisoned bees, include: *R. lutea* Sweet, *R. arboreum* Sm., *R. prattii* Franch., *R. ponticum* L., and *R. thomsonii* Hook. f.[7].

    It is worth noting that hybrids and cvs. differ in toxicity in an unpredictable manner[7]. Some spp. usually produce little N, but toxicity can be evident in P, and thus poison P foragers[7]. Typical symptoms exhibited by poisoned honeybees include, a general paralysis, curled abdomen, twitching, rapid flight from the colony only to land on the ground and spin around[68]. Rhododendrons and azaleas are cultivated or naturalized world-wide in humid and temperate regions. They are especially popular ornamentals in e. Canada and B.C.

Rhus glabra L.
smooth sumac, red sumac, scarlet sumac, vinegar tree, sumac à bois glabre
shrub   zone 2b   Jun-Jul   1.8-3 m   N
Value for honey: H yield 20-45 kg/colony/season[17]; N secreted freely on hot days, but ceases on cloudy or foggy days[17].
Honey: golden in colour; classed as amber; granulates slowly; flavour bitter when fresh, but becomes mild and rich after a few months[17].
Notes: excellent for mass planting, especially on slopes or large sites; spreads by suckering; var. *borealis* Britt.[23] is found locally in N.E. to s. Man.[23]: see Table 18.

Rhus typhina L.
staghorn sumac, velvet sumac, vinaigrier
shrub 3   Jun-Jul   3-7.5 m   N
Value for honey: HP2/3[17]; N sugar concentration 33-53%[17].
Honey: classed as amber[17].
Notes: useful for mass plantings or for screening; produces suckers; short-lived; prefers dry or gravelly soil, Gaspé Pen., Que., to s. Ont. and beyond Can. range[23].

Ribes aureum Pursh
buffalo currant, golden currant, groseillier
shrub   zone 2   Apr-May   1.8 m   NP
Value for honey: N produced abundantly[72]; corolla (blossom) length measures 1.3 cm or more[72]; in spite of the large blossom bees were observed to forage successfully by means of an unexplained slit down one side of the blossom[72].

Notes: showy fragrant flowers; sun or part shade; sometimes confused with *R. odoratum* (Missouri or buffalo currant), although the flowers of this spp. are larger.

Ribes odoratum H. Wendl.
buffalo currant, Missouri currant, groseillier des buffles
shrub   zone 2   May   180 cm   NP
Notes: yellow fragrant flowers; susceptible to black stem rust of wheat and should not be grown in wheat producing areas[81]; natural range beyond Can. limits, but has frequently spread from cultivation in B.C. and E. to N.E.[23]; sometimes confused with *R. aureum* (golden or buffalo currant).

Ribes rubrum L.
red currant, northern red currant, gadelier, groseillier rouge, raisin de mais.
shrub   Apr-May   NP
Value for honey: HP[16].
Honey: H is not usually known in pure form[35]; not commonly cultivated in N. American gardens.

Ribes sanguineum Pursh
winter flowering currant, red flowering currant, groseillier sanguin
shrub   zone 7   mid May   2.4 m   NP
Value for honey: observed to be not very attractive to honeybees; N sugar concentration 18%[79]; some ornamental cvs. have corollas (blossoms) too long for honeybees to obtain the N[35].
Value for pollen: worked especially for P[79].
Notes: ornamentally grown for its clusters of red flowers.

Ribes uva-crispa L.
(syn. *Ribes grossularia* L.; *Grossularia uva-crispa* (L.) Mill.)
English gooseberry, groseillier épineux, vinettier
shrub   -29° to -23° C   180 cm   NP
Value for honey: N is easily accessible since the bellshaped flowers are only 5 mm deep[35].
Notes: an alternate host for white pine blister rust, and cultivation may be forbidden in some areas[81]; introduced and escaped from cultivation to thickets etc.[23].

Robinia hispida L.
rose acacia, bristly locust, moss locust, acacia rose, robinier rose
shrub   zone 5   early Jun   1.5-2 m   NP
Value for honey: observed to be attractive to bees, and has been recommended for planting along roadsides to increase bee forage (i.e. N.J.)[62].
Notes: low shrubby trees with pendulous clusters of rosy-pink flowers; suckers and is useful for erosion control on banks of poor soil; found naturally in dry woods, thickets and slopes[23].

Robinia pseudoacacia L.
false acacia, black locust, yellow locust, faux acacia
tree   zone 4   early Jun   9-15 m   N(D)P
Value for honey: wide range of HP values (i.e. HP1-6)[17]; HP known to vary widely between cvs., for instance, *R. pseudoacacia* 'Decaisneana' is a cv. with a high value of

HP6[17]; HP decreases after a tree is 15 years old[17]; N sugar concentration 34-59% (max. 67%)[17]; N secretion best at temperatures greater than or equal to 27° C[17]; H yield is variable, 18-80 kg/colony/season[17]; individual flowers last for about a week, and total flowering time is up to 2 weeks[17]; honeydew produced in some years (e.g. 10-12 kg/colony/season attributed to honeydew on occasion)[17]; see Section 9.

Value for pollen: P protein content is low, but P is abundant[17].

Honey: pale yellow in colour; very clear; classed as water-white; granulation is very slow, and may take up to several years; heavy body; high fructose and low enzyme content[17].

Notes: pendulous clusters of fragrant white flowers; wide range of soils tolerated, even those which are slightly acid or droughty; is used for erosion control on slag heaps, spoil banks and roadsides in some countries[17,35]; susceptible to borer and leaf-miner pests; not sensitive to air pollution[95]; woods and thickets, naturalized N. to N.S., Que. and Ont. from points S., much planted for valuable timber and ornament[23].

Robinia pseudoacacia L. 'Semperflorens'
perpetual black locust
tree   zone 4   early Jun-Sep   9-15 m   NP

Value for honey and pollen: blooms intermittently throughout the summer instead of flowering for one short period only and therefore lacks one of the main drawbacks of the sp., *R. pseudoacacia* (black locust)[35]; has been recommended for planting on waste land and mining dumps in parts of Europe in order to improve bee pasturage[35].

Notes: similar in appearance to the sp., *R. pseudoacacia*.

Rosa (see individual spp. for more detail)
rose, brier
shrub   May-late Jun   P

Value for pollen: rose P is often greedily gathered by honeybees, and is considered an important source in areas where native roses are common[35]. Single cvs. of the shrub rose offer a timely source of abundant, nutritious P; flowers later than most other shrubs and blooms for a longer period[81]. Not all of the thousands of cvs. of garden roses (e.g. hybrid teas, polyanthas) are useful P sources since many of these have been bred to provide luxuriant bloom at the expense of the sexual parts.

Notes: Shrub roses are not as troubled by pests and disease to the same extent as other garden roses, and so usually require fewer sprays which are potentially harmful to bees.

The following few examples are chosen because they bloom at different times, and like most shrub roses, probably offer attractive P. In addition, these five native roses may be significant in local areas: *Rosa tomentosa* Sm.[23], roadside thickets in P.E.I.[23]; *R. eglanteria* L.[23] (sweet brier), thickets, roadsides and clearings throughout Can. range[23]; *R. nitida* Willd.[23], bogs, wet thickets, pond margins etc., in acid soil, Nfld. to s. Que. S. to N.S. and s. N.E.[23]; *R. johannensis* Fern.[23], gravelly shores, rocky banks, chiefly riparian, Que., N.B. and n. Me.[23]; *R. rousseauiorum* Boivin[23], marshes and borders of the sea, St. Lawrence R. and Gulf, Que.[23].

Rosa blanda Ait.
meadow rose
shrub   zone 2   early   1.8 m   P

Notes: lacks prickles; dry to moist, rocky (calcareous to neutral) slopes, shores etc., Anticosti and Mingan Ids., Que., to Man., S. to N.B. n. and w. N.E. and beyond Can. range[23].

Rosa hugonis Hemsl.
meadow rose
shrub zone 2 early Jun 1.8 m P
Value for pollen: see *Rosa.*
Notes: free flowering and vigorous; introduced.

Rosa multiflora Thunb.
Japanese rose
shrub zone 5b mid Jun 2.4 m P
Value for pollen: see *Rosa.*
Notes: profuse bloom of white flowers; hedge or erosion control; excellent for retarding sheet erosion on embankments and preventing formation of gullies[62]; when planted directly in a gully will retard further wash-outs and hold soil[62]; annual mowing will keep the plants in bounds along roadsides[62]; introduced and naturalized in clearings and borders of woods, roadsides s. N.E. and beyond Can. range[23].

Rosa rugosa Thung.
rugosa rose
shrub zone 3 early Jun 1.8-2.4 m P
Value for pollen: see *Rosa.*
Notes: flowers intermittently all summer; several single cvs. available; roadsides, seashore thickets, dune-sands, etc., Que., N.S., N.E., lower Great Lakes and beyond Can. range[23].

Rosa setigera Michx.
prairie rose
shrub zone 2 late Jul-Aug 2.4 m P
Value for pollen: see *Rosa.*
Notes: rambling type habit.

Rosa wichuraiana Crép.
memorial rose
shrub zone 5b(?) mid Jul procumbent
Value for pollen: see *Rosa.*
Notes: makes an excellent and attractive ground cover.

Rosmarinus officinalis L.
rosemary, romarin, rosmarin, herbe aux coronnes encessier
shrub zone 8 Apr-Jun 50-200 cm NP
Value for pollen: HP4[17]; N sugar concentration 25-63%[17]; H yield 60 kg/colony/season or 5-7 kg/colony/day[17].
Honey: clear; classed as water-white to white; granulation is usually rapid and fine; flavour and aroma distinctive[17]; a specialty H. The famous H of Narbonne (France) is derived largely from this plant[16].
Notes: reliably hardy only in the warmest parts of Vancouver I.[81]; fresh tops used to distill oil for perfume and medicine manufacture; leaves used in seasonings[23].

Rubus L. (*Rubus laciniatus* Willd.[51])[a]

bramble, boysenberry, loganberry, cut-leaved blackberry, murier sauvage ronce, muron, murier des haies

shrub   Jun-frost (peak in Aug)   N(D)P[17]

Value for honey: HP1/2[17]; N sugar concentration 12-47%[17]; N secretion is higher during the day following a cool night[17]; N secretion begins before flowers open, and continues until petal fall[54]; an important source of surplus in B.C.[20]; some forms are more readily visited than others[54].

Honey: light coloured; classed as white or water-white, or light amber; granulation is slow; when granulated, appears dull and waxy; flavour delicate[17,54]; sometimes H is considered inferior in quality, especially flavour, to *R. idaeus* (raspberry)[35].

Rubus deliciosus Torr.

boulder raspberry, northern raspberry, Rocky Mountain raspberry, Rocky Mountain flowering raspberry

shrub   zone 3b   May-Jun   2.1 m   NP[36]

Notes: large, solitary, white flowers; cultivated as an ornament, but natural range is restricted to canyons in Colo.[23].

Rubus idaeus L.

raspberry, European raspberry, framboisier

shrub   zone 3   Jun-Aug   1.8 m   NP

Value for honey: HP3[16]; H yield 60-130 kg/colony/season[17]; N sugar concentration up to 49% (for cultivated shrub)[17]; N flow is annual and reliable[17]; N secretion starts before flowers are open, so that bees initially forage between the petals for N[17]; N secretion is highest during the day, following a cool night, and is reduced by boron deficiency[17]; pendulous flowers protect N from dilution by rain or dew[35]; apiary locations near commercial raspberry plantations are considered to be excellent sites.

Notes: one of the parents of many modern raspberry cvs.[95]; commercial production has decreased recently due to the expense of harvesting[54].

      Many forms available; common along roadsides, in thickets and near habitations, Nfld. to Ont., S. to N.S., s.N.E., N.Y. and beyond Can. limits[23]; var. *inermis*[23] is indigenous on Magdalen Ids.[23].

Rubus occidentalis L.

wild black raspberry, black cap raspberry, thimbleberry, framboisier noir, framboisier d'Amerique

shrub   -37° to -29° C   Jun   2.7 m   NP

Value for honey: N abundant (13 mg/flower) and rich; flowering period of one plant lasts 1-3 weeks, but individual blossoms last only a day or so[54].

Value for pollen: P available all day during flowering[54].

Notes: rich thickets, ravines and borders of woods, Que. and S. beyond Can. limits[23]; fruit purplish black, but yellowish or amber in one form[23]; sometimes cultivated for fruit.

---

[a] As described in Crane et al.[17] under the syn., *Rubus* L. (*R. fruticosus* L.).

Rubus odoratus L.
flowering raspberry, thimbleberry, purple flowering raspberry, chapeaux rouges
shrub   zone 3   early Jul   2.4 m   NP[35]
Value for honey and pollen: blooms for several weeks[81].
Notes: fragrant purple flowers of ornamental interest; thickets and borders of woods, s.Que. and s.Ont., S. to N.S., N.E., L.I. and S. beyond Can. limits[23].

Rubus parviflorus Nutt.
western thimbleberry, salmonberry, thimbleberry
shrub   -37° to -29° C   1.8 m   NP
Value for honey: a well known source in some areas, including coastal B.C. in particular[72]; thickets and borders of woods, Bruce Pen., Ont., Alta. and s.Alaska, S. along Pacific coast beyond Can. limits[23]. ˙

Salix L. (see individual spp. for more detail)
willow, osier, saule
tree, shrub   Feb-late Apr   (D)NP
Value for honey and pollen: Some observers argue that willows are the most valuable plants to the apiarist[35], offering abundant P and N very early in the season. Male and female catkins (flowers) are borne on separate plants, but both offer N[35]. Honeydew, sometimes produced and collected by honeybees, is considered unsuitable for winter stores[17] (see Section 9).

In addition to those described below, the following willows are also known to be attractive to bees; *S. bebbiana* Sarg.[23](long-beaked willow, chaton or petit minou) is one of the most wide ranging and variable sp.[23], found in moist to dry thickets Nfld. to Alta., S. to N.S., N.E. and beyond Can. limits; *S. bebbiana* var. *caprefolia* Fern.[23], Nfld. and Côte Nord, Que., S. to N.S. and var. *luxurians* Fern.[23], near lower St. Lawrence R., Rimouski Co. to Gaspé Co., Que.[23]; *S. cinerea* L.[23] (gray willow, saule gris, osier cendre), introduced and spread from cultivation, N.S. and Mass.[23]; *S. fragilis* L.[23] (crack willow), spread from cultivation to roadsides, borders of woods etc., Nfld. to Ont., S. to N.S. and beyond Can. limits[23]; *S. sericea* Marsh[23] (silky willow), low thickets and banks of streams, s.Que to Wisc., N.S. N.E. beyond Can. limits[23]; *S. triandra* L.[23]; and *S.* × *smithiana* Willd.[35,36]. It is worth noting that *S. bebbiana* blooms later than most willows, just before dandelion, and honeybees are said to prefer it to dandelion or tree-fruit bloom[72].
Notes: catkins appear on the naked shoots of the previous summer's growth, so willows should be pruned only just after flowering[81]; see Table 18.

Salix alba L.
white willow, osier blanc, saule blanc
large shrub like tree   zone 4   18 m   (D)NP
Value for honey: HP4[17]; honeydew may also be collected; see *Salix.*
Value for pollen: P yield is fair[17]; P load is lemon yellow[17]; see *Salix.*
Notes: much cultivated, spread and naturalized throughout Canada[23]; see *Salix.*

Salix caprea L.
goat willow, pussy willow, florist's willow, sallow, saule des chèvres
shrub   zone 5   early Apr-mid Apr   4.5 m   (D)NP
Value for honey: HP4[17]; H yield 2-4 kg/colony/day[17]; N sugar concentration figures may vary from 15-79%[17]; honeydew may be collected[17]; see *Salix.*

Value for pollen: P yield is heavy and P is rich in protein[17]; see *Salix*.

Honey: golden-yellow in colour; classed as light amber; fine granulation; thin body; flavour mild[17].

Notes: cultivated and spread to thickets, locally, N.E., S. to beyond Can. range[23]; catkins appear before leaves.

Salix discolor Muhlenb.
pussy willow, large pussy willow, petit minou, chaton
large shrub   zone 3   mid Apr-late Apr   6 m   NP

Value for honey: H yield 6.7 kg/colony/season reported in B.C.; in an exceptionally warm season 13.5 kg/colony recorded[72]; recommended for roadside planting to increase bee forage (i.e. N.J.)[62]; see *Salix*.

Value for pollen: few early blooming plants furnish such an abundance of P[72]; see *Salix*.

Honey: said to have a pleasant, aromatic taste[72].

Notes: does best in moist soil; ideal plant for reclaiming depleted soils and retarding erosion[62]; grows in fill or mineral soil[62]; damp thickets or shores, often in swamps, Lab. to Alta., S. to Nfld., N.S., N.E.[23]; var. *overi* C.R. Ball[23] is found in Nfld. and Côte Nord, Que., to Man., S. to P.E.I., N.B. and beyond Can. range[23].

Salix nigra Marsh.
black willow, saule noir
shrub   Feb-Apr   less than 30 m   NP

Value for honey: H yield is high, up to 45 kg/colony/season[17]; see *Salix*.

Honey: is classed as extra light amber to light amber; flavour "weedy"[17].

Notes: banks of streams, shores and rich low woods, N.B., s.N.E., L.I. and beyond Can. range[23].

Salix purpurea L.
purple leaf osier, Arctic willow, basket willow, saule pourpré
shrub   zone 2b   early May-mid Apr   2.7 m   NP[36]

Value for honey and pollen: see *Salix*.

Notes: low grounds, Nfld. to Ont. and Wisc., S. to N.S., N.E., originally introduced from Europe for basket-making[23].

Senecio greyii Hook. f.
shrub   zone 9a   Jun-Jul   120 cm   N[20]

Notes: survives all but the severest winters in mildest parts of coastal B.C.; yellow daisy-like flowers and gray foliage.

Sheperdia argentea (Pursh) Nutt.
silver buffalo berry, buffalo berry
shrub   zone 1   mid Apr   4.5 m   NP

Value for honey and pollen: useful for early stimulation[35].

Notes: flowers appear in clusters on joints of stems, on the previous year's growth; useful hedge plant; resistant to drought; sexes separate; sometimes grown for edible fruit; banks of streams, Man. to Alta. and S. beyond Can. range[23].

Sophora davidii (Franch.) Skeels
(syn. *Sophora viciifolia* Hance)
vetch sophora
shrub   -23° to -21° C   Jun   2.1 m   NP
Value for honey and pollen: worked well by bees[35].
Notes: bluish-violet to whitish pea-like flowers; poor, sandy soils; belongs to the legume family.

Sophora japonica L.
Japanese pagoda tree, Japanese acacia, Chinese scholar tree, arbres des pagodes du Japon
tree   zone 6b   late Aug-late Sep   7-18 m   NP
Value for honey: HP4[16]; extremely attractive, and the "hum" of bees can be heard yards away[35]; honeybees also forage on the blossoms on the ground, which fall while still fresh[35]; apparently poor yielders after a wet cool summer[35]; trees yield only if greater than 30-40 years old[35]; produces "gum guar", rich source of a toxin, and N should be investigated for poisonous properties (although there have been no substantiated reports of honeybee or human poisonings)[7].
Notes: very long-lived; no apparent insect or disease problems; adapted to city conditions[95].

Sorbus aria (L.) Crantz.
white beam, chess apple, alisier blanc
tree   zone 4   May   4-10 m   N(P?)[35]
Notes: grows well on alkaline soils.

Sorbus aucuparia L.
(*Pyrus aucuparia* (L.) Gaertn.)
European mountain ash, rowan tree, rowan
tree   zone 3   late Jun   7-13 m   NP
Value for honey: HP2[16]; worked well, but not with enthusiasm[35].
Notes: grown chiefly for ornamental fruit and attractive foliage; roadsides, borders of woods and other habitats near towns, spread from cultivation, Nfld. to Alaska, S. to N.E., N.S., and beyond Can. limits[23]; hybridizes with *S. americana* Marsh.[23] (American mountain ash, missey mossey, cormier)[23].

Sorbus intermedia (J.F. Ehrh.) Pers.
Swedish white beam
tree   zone 4   early Jun   12 m   NP
Value for honey: freely worked for N; seems to be more attractive than *S. aucuparia* (European mountain ash).

Symphoricarpos albus (L.) S.F. Blake
(syn. *Symphoricarpos racemosus* Michx.)
snowberry, waxberry, belluaine
shrub   zone 2   Jun-Aug   1.8 m   NP
Value for honey: HP3/5[16]; flowers are said to be worked in preference to *Trifolium repens* (white clover); considered an important source in w. Canada[35,72].
Honey: is classed from white to extra light amber; flavour is mild[17].
Notes: tolerates shade; noted for terminal clusters of showy, snow white berries;

calcareous ledges, barrens and gravels, Rimouski Co., Que., to B.C., s. to w.Mass. and beyond Can. limits[23]; *S. rivularis* Suksd. (syn. *S. albus* var. *laevigatus* S.F. Blake)[23] is naturally found in roadsides, rocky banks, etc., spread from cultivation, from Que., W. and S. (introduced from the Pacific Slope)[23].

Symphoricarpos occidentalis Hook.
badger bush, wolfberry
Value for honey: H yield of 107 kg/colony/season reported in B.C. in the past (1925)[72]; N sugar concentration 57%[94].
Honey: undoubtedly dark[94].
Notes: bluffs, dry prairies and plains, Ont. to B.C., S. and E. beyond Can. limits[23].

Syringa L.
lilac, lilas
shrub   early May-late Jun   3.5-5.5 m   (N)P
Value for honey: secretes N freely, but most cvs. have corollas (flower tubes) too long for honeybees unless N is produced very abundantly and rises 2-4 mm or even more in the corolla[35]; not usually considered a valuable N source, and a single account of honeybee poisoning recorded[7].
Notes: a popular flowering shrub found in many gardens across Canada; 2 dozen spp. and over 500 cvs. available; to encourage the most profuse flowering, remove any suckers from the base of the plant; blooms on last year's wood, therefore prune after flowering; sometimes grafted on privet[95].

Tamarix gallica L.
French honeysuckle, manna plant, tamarin de France, tamarasque
shrub   -30° to -23° C   Jul-Aug(?)   1.2 m   NP
Value for honey: H yield of 45 kg/colony/season reported where this occurs naturally[72]; N sugar concentration of 37% (average)[94]; yields well in arid regions[50]; may bloom twice in milder zones[90]; said to literally "swarm" with bees, especially between legumes flows[94]; *Tamarix* H has a poor reputation for palatability (see Section 8).
Value for pollen: an abundant source of P[72].
Honey: opinions vary; dark in colour; can be classed as amber; minty[72] or "off" flavour[94] (see Section 8).
Notes: small pink flowers in racemes, 5 cm long; long lasting bloom; does well in seaside gardens; s. USA. and frequently escapes from cultivation, N. to Mass.[23].

Tamarix ramosissima Ledeb.
(syn. *Tamarix pentandra* Pall.)
amur tamarisk, five stamen tamarisk
shrub   zone 3   Jul-Aug   4.5 m   NP
Value for honey: probably similar to *T. gallica* (French tamarisk); attractive to bees[35]; *Tamarix* H has a poor reputation for palatibility (see Section 8).
Honey: probably similar to *T. gallica* H, and may have poor flavour (see Section 8).
Notes: small pink feathery flowers borne on new growth; does well near the coast; naturalized and frequently cultivated, especially in arid regions.

Taxus baccata L.
English yew, if, if d'Europe
tree   zone 7   Mar   9 m   (D?)P
Value for honey: no N[35]; honeydew H may be collected.
Value for pollen: produces an abundance of light, powdery, wind-borne P[35]; worked well by bees in some areas, but may be collected only when no other source available[35]; protein content poor[83].
Notes: excellent ornamental; sexes separate; seeds and foliage poisonous; susceptible to scale insects, so honeydew and moulds may consequently disfigure foliage; *T. canadensis* Marsh.[23] (ground hemlock, bois de sapin) is a native sp. and occurs from Nfld. to Man., S. to N.S.[23].

Tilia L.
basswood, linden, lime, lime tree, whitewood, tilleul
tree   mid Jun-mid Aug (approx.)   N(DP)
Value for honey and pollen: HP4-6[17]; most are known to be excellent N sugar producers; are commercially grown in the USSR. and China for N (and fibre)[84]; the source of 25% of the total H produced in the UK. and 75% of the total in the USSR.[84]. Unfortunately basswood (*T. americana*) stands in Canada have been dramatically decreased due to unrestricted cutting around the turn of the century. This has had a most serious and detrimental effect on the beekeeping in Ont. and Man.[16,57,84].

Mixed composition stands (i.e. hardwoods and softwoods) were found to be more reliable producers of flowers, P, and, N than single sp. plantations[84]. Fertiliser applied to roots was reported to increase numbers of flowers and N flow[84]. The amount of blossom on a tree is directly proportional to the amount of light reaching the tree (e.g. in stands with densities of 90% and 40%, trees aged 110 years have a maximum potential of 40,000 and 62,000 flowers respectively)[84]. More flowers are produced on the s. and w. part of the crown, than the n. and the e. for this reason[84]. Total N production in a stand of 90% density, was highest at 80 years and gradually decreased thereafter[84]. It has been suggested that *Tilia* could be used profitably in a forest management schemes, yielding N and timber. Mixed planting with black locust was recommended for waste land under power lines to increase bee forage[40,84].

Occasionally, *Tilia* N, P and honeydew has been reported to be poisonous to bumblebees and honeybees[7]. Signs of poisoning include, loss of ability to fly, rapid crawling, paralysed bees at the hive entrance and piles of dead bees below the trees[7]. Blossom N and honeydew proved to be toxic when fed to bees, but are harmless to most mammals[7]. Only when bees visit *Tilia* exclusively, is there a noticeable decrease in colony populations[7]. Bees tend to be most severely affected in dry years[35]. Reputations of *Tilia* in connection with reports of poisonings vary, and *T. petiolaris* (pendant silver linden) has one of the worst reputations among beekeepers[55].

Caution should be exercised when blaming certain *Tilia* as the sole cause of bee poisonings, because of the inconsistency of the reporting and confusion between spp. due to incorrect naming[a]; see Sections 9 and 10.

Table 5 shows how the total flower time of lime can be extended if several *Tilia* are grown together. Although the actual dates will vary according to the area, the flowering sequence tends to remain the same. Whenever possible, spp. known to be less prone to aphid infestation should be grown, in order to reduce the amount of honeydew produced (a nuisance in urban areas) or collected in the crop (usually undesirable in

---

[a]   Patch, D. (1983). *Personal communication* .

honey).

TABLE 5. Approximate flowering times of selected basswood (*Tilia*) (from Melville[55]).
* indicates that a species usually produces less honeydew than other *Tilia*

Species	Period of flowering						
	June			July			August
	10	20	30	10	20	30	9
T. platyphyllos Scop.		————					
T. platyphyllos Scop. 'Asplenifolia'		———					
T. × europaea L.		———					
T. americana L.		———––					
*T. maximowicziana Shiras.			——–––				
*T. mongolica Maxim.			——––––				
*T. × euchlora C. Koch				——––––			
T. cordata Mill.				————			
*T. tomentosa Moench				——–––			
T. orbicularis					———––		
T. petiolaris DC.					——––		

T. maximowicziana has been observed to be more attractive than other commonly grown spp.[35]; *T. amurensis* Rupr. (HP6[17]) and *T. japonica* (Miq.) Simonk. (H yield 60 kg/colony/season[17]) are both favoured by bees. However, since these three spp. are rare in Canada, they are not described below.
Honey: light, somewhat greenish; granulates rapidly and smoothly; pronounced characteristic flavour and aroma; density is often low[17]; see Section 9.
Notes: one of the major hardwood spp. in the deciduous forest region of s.Ont.[84]; an important group of shade trees for specimen or avenue planting (valued chiefly for their foliage and habit); flowers are extremely fragrant; wood is highly valued because it is light coloured and clear.
The following are worth considering for street planting; *T. × euchlora* (Crimean linden); *T. cordata* (little leaf linden); *T. mongolica* (Mongolian linden); *T. platyphyllos* (big leaf linden) and cvs.

Tilia americana L.
basswood, American lime, whitewood, bois blanc
tree   zone 2b   late Jun   18-20 m   N(P)
Value for honey: H yield is up to 22 kg/colony/season[17]; N flow is irregular, usually 2-3 good years out of 5[17]; N flow usually lasts for 10-14 days and is stopped by rain and temperatures below 18° C[17]; H yield 9.1 kg/colony/day has been measured[80]; see *Tilia* .
Honey: classed as water-white to white[17].
Notes: large coarse tree, not often grown ornamentally; harvested commercially for timber or fibre in Canada; the wood is often used in beekeeping industry for making comb honey sections[72]; found in rich woods s.Que. to Man., S. to N.B., N.E. and beyond Can. limits[23]; see *Tilia*.

Tilia cordata Mill.
(syn. *Tilia parviflora* J.F. Ehrh. ex Hoffm.)
small leafed linden, small leaved European linden, tilleul à petites feuilles
tree   zone 3   early Jul   12-15 m   N(D,P)
Value for honey: HP4/6[16]; H yield 15-20 kg/colony/season[17]; H yield per tree up to 30 kg[17]; N sugar concentration 17-56%[17]; N flow usually lasts for 12 days; flowers not pendant (like other limes) and N may be diluted by rain[35]; honeydew collected only in early morning or late in the day (and/or when RH is high) because at other times it granulates on the tree and is difficult for the bees to collect[17]; bees may show signs of poisoning in some years (especially under drought conditions)[17](see Sections 9 and 10).
        The greatest concentration of this sp. in the world (1.1 million ha) is found in mixed stands in the Middle Volga Region, USSR., a very important beekeeping area[84].
Value for pollen: yield is low; P may be discarded by bees; see *Tilia* .
Honey: water-white in colour (honeydew content alters this to dark green); granulates slowly; flavour is reminiscent of peppermint, with a strong aroma of lime[17].
Notes: one of the best *Tilia* for street and avenue planting[95]; see *Tilia*.

Tilia × *euchlora* C. Koch
Crimean lime
tree   zone 3   early Jul   15 m   NP(D)
Value for honey: not as freely worked as some other limes[35]; see *Tilia*.
Honey: very little honeydew collected[35]; see *Tilia* and Section 9.
Notes: remarkably free of insect and disease pests, and honeydew[35]; considered as an ideal avenue or street tree[35]; one of the most beautiful of the lindens[35]; see *Tilia*.

Tilia × *europaea* L.
common lime, lime, common linden, tilleul des bois
tree   -37° to -29° C   mid Jun   36 m   N(D,P)
Value for honey: N flow is 2-3 weeks[35]; N sugar concentration 32-71%[17]; sometimes a source of surplus H in Charlottetown, P.E.I.[72]; see *Tilia* .
Honey: prone to yield honeydew H[35]; see *Tilia* and Section 9.
Notes: no longer recommended as an ornamental tree because of its profuse suckering and susceptibility to plant lice[95]; see *Tilia*.

Tilia mongolica Maxim.
Mongolian linden
tree   zone 3b   mid-Jul   13 m   N(P)
Value for honey: freely worked by bees[35]; see *Tilia* .
Honey: little honeydew H; see *Tilia*.
Notes: very free flowering; handsome small-avenue or street tree[95]; see *Tilia*.

Tilia platyphyllos Scop.
big leaf linden, large leaf linden, tilleul à grandes feuilles
tree   zone 4   mid Jun   15-25 m   N(D,P)
Value for honey: HP4/6[17]; H yield 15-20 kg/colony/season[17]; N sugar concentration 21-48%[17]; N flow lasts for 10 days to 3 weeks[17]; optimum temperature is 18° C-21° C for N secretion[17]; N flow is greater with warm nights, warm days, and high RH[17]; N, P, or honeydew may occasionally be responsible for poisoning bees[7]; *T. platyphyllos* 'Asplenifolia' is worked well by bees[17]; see *Tilia* and Section 10.
Value for pollen: P yield is low[17].

Honey: honeydew H is dark green in colour; granulation is very slow; see *Tilia* and Section 9.
Notes: a coarse tree, but ornamentally more desirable than *T. × europaea* (common lime); see *Tilia* .

Tilia tomentosa Moench
silver leaved linden, silver linden, tilleul argenté
tree   zone 6   early Jul   12-18 m   N(P,D)
Value for honey: HP6[17]; N flow lasts for about 8 days[17]; N sugar concentration 26-63%[17]; N, P or honeydew may occasionally poison foraging bees[7]; see *Tilia*.
Value for pollen: see *Tilia*.
Honey: see *Tilia*
Notes: named for the white, soft underside of the leaves that give the whole tree a whitish appearance, especially in a slight breeze; excellent specimen tree; see *Tilia*.

Ulex europaeus L.
gorse, common gorse, furze, whin, ajonc, jonc marin
Value for honey: does not produce N[a].
Value for pollen: a good shrub for seashore planting; naturalized on Vancouver I., B.C.; sometimes cultivated as a sand binder, now locally naturalized on sands from se.Mass.[23]; yellow pea-like flowers which have a second bloom in late summer.

Ulmus americana L.
American elm, white elm, water elm, orme
tree   zone 7   early Apr   18-36 m   (N)P
Value for honey: some elms may supply N and honeydew[72].
Value for pollen: an abundant supply of windborne P, eagerly collected, especially in the proximity of hives[72]; P has a good protein content[83].
Notes: large well-known ornamental tree; rich soil, especially along streams or in lowlands, Gaspé Pen., Que. to Sask., S. to N.S., N.E. and beyond Can. limits[23].

Vaccinium L.
blueberry, cranberry, bilberry, huckleberry, airelle
shrub   May-Jun   NP
Value for honey: There is considerable variation in the attraction of certain cvs. of blueberries and cranberries[54]. The following are not described here, but have been observed to be attractive to honeybees: *V. uliginosum* L.[23] (black whortleberry, bog bilberry, grandes brimbelles des marais), HP4/5[17]; and *V. myrtillus* L. (bilberry, blueberry, common whortleberry, airelle, myrtille noir, raisin de bois)[72].
Honey: white in colour; may granulate hard; good body; flavour mild[16].
Notes: interesting and useful shrubs of particular merit for the beauty and texture of their leaves; low growing, wet areas; some spp. are tended in wild or garden plantations for commercial fruit production.

---

[a] Adey, M. (1986). *Personal communication.*

Vaccinium angustifolium Ait.
(syn. *Vaccinium pensylvanicum* Lam.)[51]
low bush blueberry
shrub   -50° to -37° C   May   20 cm   NP
Value for honey: H yield 7-16 kg/colony/season[50]; some cvs. are much more attractive to bees than others[54].
Honey: see *Vaccinium.*
Notes: cross pollination is accepted as being essential[54]; native bee populations are not usually dependable, and recommended rates for pollination service vary from less than 1, up to 2 colonies/ha[54]; in N.S., less than 1 colony/ac (2 1/2/ha) increased fruit production 2-4 times that where no colonies were present[72]; see Table 18.

Vaccinium arboreum Marsh.[51]
tree huckleberry, farkleberry, sparkleberry, winter huckleberry, tree berry
shrub   -15° to -12° C   Jun   8 m   N(P?)
Honey: amber in colour, with a reddish tinge; flavour strong[72].
Notes: sometimes grown as an ornamental, but hardy only in a few parts of Canada; sandy woods, thickets and clearings, but common only in areas beyond Can. limits[23].

Vaccinium corymbosum L.
highbush blueberry, whortleberry, swamp blueberry, northern highbush blueberry, airelle à corymbes
shrub   zone 4   late May-early Jun   3.6 m   NP
Value for honey: H yields of 22.5-40.5 kg/colony/season reported[72]: N flow usually lasts 8-10 days[72]; N secretion may be so great that unless strong foraging force present, cross pollination may not be effected although N is obtained[72]; some cvs. are more attractive to bees than others[54].
Honey: see *Vaccinium.*
Notes: Cross pollination is accepted as essential on this crop. Native bee populations are even less dependable for the pollination of this crop than for *V. angustifolium*, low bush blueberry, due to the intense cultivation practices that are required. As a result, the extent of suitable habitat for native pollinators is drastically reduced[54].

One of the most ornamental of this genus, frequently seen in gardens; needs acid soil; swamps, low woods or even dry upland, N.S. to s.Que., W. to Wisc. and S. beyond Can. range[23].

Vaccinium macrocarpon Ait.
cranberry, atocas, large cranberry, American cranberry, canneberge à gros fruits
shrub   -50° to -37° C   May   low creeping   (N)P
Value for honey: blossoms secrete little N, and in some locations, practically none; not highly attractive to bees[54]; cold injury further reduces attractiveness and N production[54].
Value for pollen: produces a generous amount of P[54]; P is usually attractive to honeybees, but some lines of honeybees may have a heritable tendency to collect more than other lines when P collecting from this crop[54]; honeybee pollination is considered essential for economic fruit set because native pollinators cannot be depended upon in most areas[54].
Honey: light amber in colour with fine flavour, but is seldom obtained[54].
Notes: open bogs, swamps and wet shores, Nfld., S. to N.S., and beyond Can. range[23].

Vaccinium vitis-idaea L. var. *minus* Lodd.
creeping mountain cranberry, mountain cranberry, cowberry, lingen, lingberry, red whortleberry, "pomme de terre", airelle rouge, myrtille rouge
shrub   zone 1   late May   20 cm   N(P?)
Notes: a creeping ground cover; rocky or dry peaty, acid soil, subarct. America, S. to Nfld., N.S., L. Sup. reg. of Ont., Man., Sask., Alta., and B.C.[23]; edible fruit.

Viburnum opulus L.
European high bush cranberry, Guelder rose, Whitten rose, rose de Gueldre
shrub   zone 2b   early Jun   3.6 m   N[36]
Notes: vigorous shrub with flat clusters of white flowers; frequently planted and sometimes escaping n.'ward[23].

Viburnum plicatum Thunb. cv. *tomentosum*
Japanese snowball
shrub   zone 6   late May   2.7 m   NP[36]
Notes: large conspicuous blooms.

Viburnum prunifolium L.
black haw, sweet haw, nannyberry, stagbush, viorne à feuilles de prunier
shrub   zone 3   late May   5.4 m   N(P)[72]
Notes: thickets and borders of woods N. to N.Y. and may occur in e. Canada[23]; one of the best of the large *Viburnum*.

Vitex negundo L. cv. *heterophylla* (Franch.) Rehd.
(syn. *Vitex negundo* L. var. *incisa* (Lam.) C.B. Clarke; *V. incisa* Lam.)
cut leaved chaste tree, gattelier
shrub like tree   zone 5   Sep-Oct   3.6 m   N(P?)
Value for honey: HP6[16]; has been described as the best yielding shrub known in some areas[72].
Honey: white in colour, with a greenish tinge[90].
Notes: graceful with an abundance of violet blue flowers; root hardy only at Ottawa[81].

Weigela florida (Bunge) A. DC.
weigela
shrub   zone 4   May-Jun   1.5 m   N[36]
Value for honey: secretes N freely, but N may be collected by honeybees using "robbing" holes, previously cut by bumblebees in the base of the blossoms[35].
Notes: blooms profusely and needs renewal pruning to keep these flowering; may kill to ground in a severe winter[81].

Wisteria floribunda (Willd.) DC.
Japanese wisteria, glycine du Japon
vine   zone 6   late May-early Jun   NP
Value for honey: attractive[72].
Notes: pendulous trusses of fragrant blossoms that open progressively toward the tip[35]; needs shelter and a south facing wall; open woods, old homesteads, etc., local, Mass., introduced and naturalized[23].

Wisteria sinensis (Sims) Sweet
(syn. *Wisteria chinensis* DC.)
Chinese wisteria, glycine de Chine
vine   zone 6b   mid May-late May   less than 30 m   NP[36]
Value for honey: attractive[72].
Notes: trusses of fragrant blossoms which open all at once[35]; open woods, roadsides and abandoned gardens, s. N.E. and beyond Can. limits[23].

Yucca filamentosa L.
Adam's needle, needle palm, silk grass, yucca, yucca filamenteux
shrub   zone 4   Jul   1.2 m   P[36]
Notes: grows best in a hot dry location; large spikes of creamy white flowers; dry sands and beaches, dunes, old fields and pinelands to s. N.J., but often cultivated and established n'ward.[23].

Yucca glauca Nutt.
soapweed
shrub   zone 3   Jul   60 cm   (N)P[72]
Notes: large spikes of creamy-white flowers; requires a hot dry location; native to Alta.; dry plains and sandhills.

6. WAYS TO IMPROVE THE ATTRACTIVENESS OF PLANTINGS TO BEES

Massing and spacing: A beekeeper or gardener who decides to go to the trouble and expense of growing plants for bees will be better rewarded if the annual and perennial flowers are massed, and the trees or shrubs planted in avenues or groups. Plants grown in this way are more pleasing to the eye because of the dramatic impact of a large splash of colour, and are also more enticing to foraging bees. Individual plants blooming in isolation are likely to be overlooked, and the reward they offer to honeybees may not "pay" for the energy it would take a bee to investigate such a small spot of colour. In addition, a mass planting is much more likely to lure foragers by scent than a single flower.

However, care should be taken not to crowd individual plants when creating a mass of bloom. Sufficient room should be given to each plant, especially trees and shrubs, to allow the plants to attain their natural width.

"Dead-heading": Although it may seem very tedious, dead flower blossoms should be conscientiously removed to encourage the plant to keep blooming. This especially applies to annual flowers and some shrubs. "Dead-heading" can sometimes extend the flowering period for several weeks.

Fertilizers: Fertilizing is often beneficial, but excessive vegetative growth, promoted by an abundance of nitrogen, is usually detrimental to nectar production. Although it has been suggested that some nutrients promote nectar secretion, it is always safest to have garden soil tested before applying any fertilizer. In this way over-fertilizing can be avoided. Excessive amounts of any one nutrient can seriously damage plant health, and thereby hinder nectar secretion. Undoubtedly, healthy, well-nourished plants will produce good blooms, and hopefully provide the anticipated nectar flow.

Pruning: If annual and some later-flowering biennial and perennial flowers (not shrubs) are pinched back several times during their early growth, branching will be encouraged, more flowers will be produced and the plant will have a better overall appearance. The final result will be attractive flower beds with an even spread of bloom.

Colour or scent: Scent serves as an important close-range guide for bees, but is secondary to colour in initially attracting them. The colours that bees are most strongly attracted to are yellow, blue-green to blue, mauve, purple and red containing ultra-violet[a]. On the whole, yellow seems to be the most appealing colour to bees[27]. This last fact is something that the beekeeper could bear in mind when considering his or her bee-yard attire. When working bees or gardening in the vicinity of an apiary it is best to avoid wearing yellow or orange so as not to be pestered by soliciting foragers[b].

---

[a]   Reds without ultra-violet are probably seen as khaki-green.

[b]   Brehaut, K. (1986). *Personal communication.*

7. PLANNING FOR CONTINUOUS BLOOM

A garden that offers the delights of colour, scent and form throughout the growing season is the ambition of many an ardent gardener, but success requires careful planning and dedication. If one is prepared to be patient, collections can be gradually increased over many years as gaps in the blooming sequence are noted. However, many of us would like to achieve instant results, with full borders in one or two seasons.

A garden plan is advisable for this approach because it is a convenient way of visualizing the final effect of a proposed arrangement, and it will also help in remembering the names of plants placed in the border. The lists in Section 13.4 could be used to select plants for a garden design according to their season of first bloom. A mixture of, chiefly, perennial flowers, some shrubs, a few annuals and biennials, and a scattering of spring bulbs are the basic ingredients of a mixed border.

In the long term, maintenance of perennial flowers and shrubs is easier and less costly than a garden dominated by annuals and biennials which have to be renewed each year. However, the annual and biennial flowers are very important as fillers in any flower border, and are usually not out of place in a garden scheme. Annual flowers can also be used to advantage in a recently planted perennial or shrub border to fill in the gaps between the long-term plants until these achieve their full size and spread.

In most of Canada, because of a severe winter climate, better over-wintering success will be achieved if the flower beds are slightly raised and contoured to improve drainage.

Clematis virginiana

Rudbekia laciniata

Ceanothus americanus

Satureja vulgaris

Liastris spicata

8. UNPALATABLE HONEY SOURCES

There is no doubt that a particular honey will taste and smell differently to each of us. Also, individual preferences vary, and what is unpalatable to one person may be delicious to another. For example some people have a palate for a strong honey, such as buckwheat, and others may prefer a very mild type of clover honey. The average North American consumer selects mild, light coloured packs, while the European and British market buys more of the strongly flavoured, specialty honeys.

However there is nothing to suggest that bees find so-called "unpalatable" honeys objectionable, nor are such honeys detrimental in any way to bees *because* of their apparent ill-flavour[a].

8.1 Floral sources

An objective definition of what constitutes a good or poor tasting honey is impossible because taste cannot be measured objectively, like water content or colour. However, some floral sources do have a well-founded and wide reputation as producers of very bad or bitter honey. Apiary sites dominated by these floral sources are usually avoided by beekeepers, because it is known to be possible to spoil a whole crop of table honey by the addition of an unpalatable honey to the surplus.

Fortunately there are few plants in Canada with this poor reputation. Five widely recognized potential sources of ill-tasting floral honey in Canada are mentioned below. In addition, other naturally occurring or cultivated plants in Canada are probably reported locally as unpalatable sources, but are not widely recognized as such.

1. *Senecio jacobaea* L. (ragwort, tansy ragwort, stinking Willie, seneçon) is a weed of fields, pastures and roadsides, that grows to be about 1 m tall. In most areas it grows as a biennial, flowering only in the second year after seeding, but sometimes it persists as a perennial. Ragwort is widespread in certain isolated localities of Newfoundland, the Gaspé Peninsula, Quebec, and throughout much of the Maritimes, especially Nova Scotia[23]. It is a casual, but increasingly serious, weed in Ontario and is established in Manitoba, Saskatchewan, Alberta and British Columbia. Ragwort is common in poor and neglected pastures, but may be found in a wide variety of situations. The plant is poisonous to livestock, and is usually avoided by grazing animals.

In some years, ragwort plants are particularly prevalent. Since it withstands drought better than clover (*Trifolium*), bees are prone to work ragwort more intensively in dry years, and the result may be noticeable in the honey crop[35].

Ragwort honey is deep yellow in colour and has a strong flavour and aroma, and is considered nauseous by some people[35].

2. *Ligustrum* (privet) is commonly grown as an ornamental clipped hedge. It can be found in most urban or suburban regions of Canada. Overgrown, neglected hedges produce an abundance of blossom, rich in nectar. Honeybees usually find privet an

---

[a] Although a nasty tasting honey might *coincidentally* be harmful to bees.

irresistible source. However, there is usually not enough nectar collected to spoil the main crop.

The pure honey is dark, fairly thick and distinctly bitter. It is said to taint other honeys if mixed with them in any quantity[35].

3. *Helenium tenuifolium* (bitterweed, orange sneezeweed) produces a canary-yellow coloured honey, which is said to be about as "offensive to the palate as quinine"[72].

Fortunately, this plant probably is not hardy in most parts of Canada. If it does occur, it is hardy only in the mildest regions of B.C. A relative *H. autumnale* (northern sneezeweed) is common in southern Canada. This species apparently has less bitter honey and is usually considered a valuable honey source by beekeepers[3,72]. However, it is also contended that a substantial surplus of honey from northern sneezeweed could spoil a honey crop[72].

4. *Magnolia grandiflora* (southern magnolia) is hardy only in the mildest parts of B.C. The honey is reported to be strong and unpalatable[72].

5. *Tamarix* (tamarisk, salt cedar) is common in the arid southwest of the USA, and is reported to yield "a poorly flavoured honey that often damages other honeys"[50]. It is unclear whether or not this reputation applies to all members of this genus, including the ornamental forms often grown in Canada (e.g. *T. ramosissima*, Amur tamarisk; *T. parviflora*, small-flowered tamarisk).

8.2 Non-floral sources

Bees sometimes gather sweet substances other than nectar, and the effect on the honey crop is not necessarily detrimental. Honeydew, the excretion of plant-sucking insects, is a non-floral source of sugar often gathered by honeybees.

In North America, honeydew honeys are generally reputed to be unpalatable (and unsuitable for winter stores). However, this reputation is not always justified and it would be misleading to discuss honeydew honeys under the heading of unpalatable. Although it is likely that certain Canadian honeydew sources do produce an ill-tasting or even unpalatable honey, there is a lack of detailed information specific to Canada. For these reasons, honeydew is covered separately, in Section 9.

Bees will attempt to collect a wide range of other, non-floral sweet substances if these are sufficiently sweet and abundant compared to other available forage. Many of these non-floral sources are associated with wastes from manufacturing and agricultural industries, but are usually not collected in sufficient quantity to be noticeable in the surplus honey crop.

9. HONEYDEW

9.1 Honeydew honey surplus

Honeydew flows are not noted as a major source of surplus except in perhaps one or two isolated situations (e.g. *Pseudotsuga menziesii* var. *glauca* , blue Douglas fir, in the Interior of B.C.). Nonetheless honeydew flows probably make an unappreciated but significant contribution to surplus in some areas of the country. Beekeepers should cautiously assume that lack of information specific to Canadian honeydew sources only indicates poor documentation, and does not really reflect the importance of Canadian honeydew to the beekeeping industry.

In some parts of the world honeydew flows are the main source of marketable and surplus honey but are very localized in nature, and therefore only a small number of beekeepers are dependent on any particular flow. The major sources of honeydew are trees in temperate zone forests (see Tables 7 and 8), especially in New Zealand, North America and parts of Europe (e.g. the Black Forest in Germany, parts of Greece, and the eastern Mediterranean).

9.2 The production of honeydew

Honeydew is the name given to the sugary substance excreted by certain plant sucking insects (Hemiptera) which have mouth-parts capable of piecing some part of the host plant (see Table 6). Such a large quantity of food is ingested (intake-per-hour can exceed body weight) that honeydew production can be very prolific. The excretion contains both insect-digested plant saps and a large proportion of sap that has only been filtered after ingestion. The honeydew, secreted in a "rain" of droplets, falls onto the surfaces of leaves and twigs and onto the ground beneath the trees. The sticky droplets solidify quickly through evaporation, and may give leaves a shiny, "varnished" appearance.

The significance of a honeydew flow to beekeepers depends on the size of the honeydew producing insect population. Any particular flow is unique to the insect species feeding on a certain part of the host tree. In some cases one tree may be host to several species of insect, and thereby is the source of several honeydew flows. Honeydew flows are sometimes initiated by beekeepers who introduce feeding insects on branches of the host tree into regions where these insects previously were absent, for example in parts of Turkey. However, the seasonal growth of honeydew producing insects is very sensitive to environmental factors such as predators, humidity, temperature and degree of atmospheric pollution (especially "acid rain"). In most cases these factors are not easy to control, and their influence cannot be predicted accurately. Garnering honey crops from honeydew flows is not without risk or challenge to the beekeeper.

Many crop and garden plants are susceptible to infestations of aphid or scale insects and thereby are a source of honeydew, sometimes collected by honeybees. These flows from cultivated plants are not significant, except in the case of unsprayed beans, peas and cereal crops. However, if pest management is effective, no honeydew will be produced to benefit the beekeeper. In any case, it is hazardous for beekeepers to exploit any crop likely to be sprayed, since pesticides could be toxic, if not lethal, to foraging bees.

TABLE 6. Some typical honeydew producing insects, commonly known as scale insects, mealy bugs or aphids (adapted from Crane *et al*[18]).

Species	Family
Cinara confinus	Lachnidae
Cinara laricis	Lachnidae
Cinara pectinatae	Lachnidae
Cinara piceae	Lachnidae
Cinara pilicornis	Lachnidae
Lachnus roboris	Lachnidae
Phyllapis fagi	Callaphididae
Physokermes hemicryphus	Coccidae
Ptreocomma salias	Aphididae
Xylococcus macrocarpi	Coccidae

The following plants (Table 7) primarily are nectar sources (they are listed in the Indexes).

TABLE 7. Well-known nectar sources that also produce honeydew (adapted from Crane *et al*[17]).

Species	Common name
Acer platanoides	Norway maple
Acer pseudoplatanus	sycamore
Helianthus annuus	common sunflower
Liriodendron tulipifera	tulip tree
Malus domestica	apple
Medicago sativa	alfalfa
Robinia pseudoacacia	false acacia, black locust
Rubus spp.	blackberry, bramble, raspberry
Salix alba	white willow
Salix caprea	goat willow, pussy willow
Tilia cordata	small-leafed linden
Tilia platyphyllos	big leaf linden
Tilia tomentosa	silver-leafed linden
T. × europaea	common lime
Trifolium pratense	red clover
Vicia faba	broad bean, field bean

The following plants (Table 8) primarily are honeydew sources rather than nectar sources (they are listed in the Indexes).

TABLE 8. Well-known honeydew sources.

Species	Common name
Abies alba	silver fir
Calocedrus decurrens	California incense cedar
Fagus sylvatica	European beech
Larix decidua	European larch
Picea abies	Norway spruce
Pinus sylvestris	Scots pine
Populus	poplar, aspen
Pseudotsuga menziesii var. *glauca*	blue Douglas fir
Quercus robur	common oak
Zea mays	corn, maize

9.3 The nature of honeydew and honeydew honeys

Honeydew[a] is *chemically* distinguishable from the original plant sap, and honeydew honey is distinctly different, *chemically* and *physically*, from floral honey. Honeydew honeys characteristically contain high amounts of nitrogen (mostly amino acids and amides), organic acids, trisaccharide sugars (i.e. melizitose, fructomaltose), dextrin and some enzymes not normally found in honey.

The main sugars in honeydew are almost always glucose and fructose. In certain cases honeydew sugars are replaced by sugar alcohols, rendering them unattractive to honeybees. Occasionally complex sugars which are harmful to bees are found in honeydew, such as raffinose, mannose, and melibose.

The trisaccharide sugar, melizitose, can be present in large amounts in some honeydews and honeydew honeys, causing the honeydew or honey to crystallize on the tree or in the comb. The highly insoluble nature of this sugar means that bees are only able to collect it when it is liquid, in the morning, or when the humidity is sufficiently high. Severe colony losses often have occurred when bees have been over-wintered on stores containing honey with a significant melizitose content. On the other hand, several other honeydew honeys crystallize slowly, if at all. However, even if the honey remains liquid, poor over-wintering results are usually expected when bees are dependent on stores of honeydew. Over-wintering losses are usually attributed to the "gum" or dextrin content in the honeydew honey.

Algal or fungal moulds may be present in honeydew, and subsequently in the honey, imparting to both a light green to dark green or brown colour. This characteristic of a honeydew honey, in addition to its distinct flavour, is appreciated on the European market, but not usually by most North American consumers who consider honeydew honeys inferior to floral honeys, or even unpalatable.

---

[a] The information in this Section (9.3) comes from the following references: 17, 18, 29, 35, 72.

10. POISONOUS PLANTS

No discussion of nectar and pollen sources can be complete without some mention of toxic or poisonous honey plants. These are plants whose nectar, pollen, honeydew or sap is toxic to honeybees, or those plants which produce honey toxic to humans or other animals. Although there are relatively few such plants in North America, the prudent beekeeper should try to be aware of these plants, and avoid placing colonies in areas where they are abundant[6].

10.1 Human poisoning by honey or pollen

There have been very few, if any, substantiated incidents where ingestion of toxic nectar, pollen or honey has caused human death. Typical symptoms reported include numbness, dizziness, nausea, and temporary loss of voluntary muscle control[43]. In the few cases which have resulted in a fatality from poisoning, death was attributed to respiratory paralysis[43]. One of the active substances implicated, andromedotoxin-acetylandomedol, has been detected in the nectar of several members of the Ericaceae or heather family (e.g. *Rhododendron ponticum*, *Kalmia latifolia*, and *Ledum palustre*)[7]. Several other Ericaceae are suspected of producing a toxic honey (e.g. certain spp. of *Pieris*, *Andromeda*, and *Leucothoe*). There is evidence to suggest that only the unripe or uncapped honey is to blame in some cases.

Table 9 gives the names of native, naturalized or cultivated plants in North America which may be responsible for honey that under certain circumstances, or in some seasons, proves harmful to humans.

10.2 Honeybee poisoning by nectar, pollen or honeydew

Although pesticide poisoning is a much more serious problem than plant poisoning, toxic plants can cause severe losses in limited areas. It is curious that some recommended honey plants (e.g. *Euphorbia marginata*, some *Pieris* spp., some *Tilia* spp., and some *Andromeda* spp.) are sometimes associated with bee poisoning. In several cases plants which are toxic to livestock, for instance *Solanum nigrum* (deadly nightshade), are also poisonous to honeybees[7].

The available data on incidents of honeybee poisoning are not entirely satisfactory. The literature consists of largely anecdotal, rather than experimentally substantiated, accounts. Emphasis has been placed on post-mortem examination of bees, and diagnostic keys have been developed to determine if death was due to "pollen poisoning", "honeydew poisoning" or "nectar poisoning". In the majority of cases these are not separate effects with a single cause, but result from a complex of biological conditions inside the hive, as well as local environmental influences[7]. Such factors have been largely neglected by researchers and other observers[7]. Abnormal climatic or soil conditions while plants are blooming may influence plant physiology, and as a result the balance of chemicals in the nectar, pollen or sap may be altered to such an extent that these substances may become toxic to honeybees[7]. This toxicity has been observed in some plants which are stressed by lack of water or a sudden drop in temperature (for instance, chilling for 6 hours causes *Crocus* pollen to become toxic[7]).

Phyto-hormones, and other chemicals which are designed to alter the sugar content of plants, have often been used by agriculturalists without enough consideration of their effects on nectars[7]. One group of sugars in particular, galactosides, are extremely poisonous to bees (i.e. 2-4% galactosides fed in 50% sugar syrup kills bees)[7], but are sometimes manufactured by plants and can be found in nectar.

In general the symptoms of honeybee poisoning by nectar are usually limited to the blooming period of the plant in question, and, if the colony survives, the symptoms disappear when the plant stops flowering. In contrast to the limited period associated with nectar toxicity, pollen toxicity can be effective for as long as the pollen remains in the combs, either in stored surplus honey or as stored pollen pellets[18]. The symptoms of pollen toxicity are sometimes mistaken for symptoms of honeybee disease. Unlike pesticide poisoning the symptoms of nectar or pollen poisoning develop more gradually, and persist for a longer time. They may also recur in the same area at similar periods each year, but not necessarily with equal severity.

However, contrary to popular opinion, danger to foraging honeybees from plant toxins is not limited to the blooming period. Sources of plant toxins may be attractive and available to honeybees before, after, and during flowering. These sources include extra-floral nectar; sap and gum exudates from injuries or ripe fruits; honeydew secreted by sucking insects[a] ; and drinking-water contaminated by plant leachates[7].

Fortunately the majority of plants which poison bees are usually those which produce little nectar and pollen, and hence are less attractive than non-toxic species. Under unfavourable circumstances, such as a drought, toxic plants may be visited more frequently than usual, and poisons are then more likely to accumulate in honey, pollen stores or a forager's honeysac. It is curious that honeybees are particularly susceptible to certain sugars, saponins, and gums which are very commonly found in plants.

Table 9 lists some plants which are reported to have been toxic to bees at some time. As some of these occur naturally, or are cultivated in Canada, beekeepers should try to take note of them and avoid placing colonies in the vicinity of any bee forage known to be potentially poisonous.

---

[a] Refer to Section 9 for further information on the toxic properties of some honeydews to honeybees.

POISONOUS PLANTS

TABLE 9. Plants described as poisonous to honeybees, or producing honey poisonous to humans or other mammals. In some cases, plants are lethally poisonous (adapted from Alex *et al.*[3], Atkins[6], Barker *et al.*, Howes[34], Leach[48], Morton[64], Oldszowy[68]).
N, toxic nectar; D, toxic honeydew; P, toxic pollen; H, toxic honey;
?, toxicity only suspected or not well-documented; x, affirmative;
*, see index for more information; occ., occasionally.

Names of plants	Naturalized in North America	Native to North America	Cultivated in Canada	Honey toxic to humans	Plant poisonous to mammals	Plant poisonous to bees
Abies alba silver fir	–	–	x	–	–	x D occ.
Aconitum monkshood	–	some spp.	x	x?	x	x P
Aesclepias milkweed	–	x	occ.	–	x[a]	x[b]
Aesculus californica California buckeye	–	x	–	–	–	x NPH
Andromeda andromeda	–	x	x	x?	–	x? HN
Arabis glabra tower mustard	–	x	–	–	–	x?
Calluna vulgaris heather	x	–	x	–	–	x? one report
Caltha palustris marsh marigold	–	x	occ.	–	–	x P occ.
Cuscuta dodder	–	x	–	–	x?	x
Cyrilla racemiflora southern leatherwood	–	x	–	–	–	x
Daphne mezereum February daphne	x	–	x	–	x	x? one report
Datura stromonium datura, Jimson weed	–	x	–	x	x	–
Delphinium consoida forking larkspur	–	x	–	–	x	x
Digitalis purpurea foxglove	x	–	x	–	x	x P
Dipsacus sativus fuller's teasel	x	–	x	–	–	x? one report
Euphorbia marginata snow on the mountain	–	x	x	–	–	x

TABLE 9. Continued

Names of plants	Naturalized in North America	Native to North America	Cultivated in Canada	Honey toxic to humans	Plant poisonous to mammals	Plant poisonous to bees
Gelsemium sempervirens yellow jasmine	–	x	occ. B.C.	x	x?	x H?P?
Helenium hoopesii orange sneezeweed	–	x	–	–	x	x
Hyoscyamus niger black hen bane	x	–	x	–	–	x
Hypericum perforatum St John's wort	–	x	–	–	x[c]	–
Juglans mandshurica Manchurian walnut	–	–	–	–	–	x?
Kalmia latifolia mountain laurel	–	x	x	x	x	x?
Ledum palustre wild rosemary	–	–	–	x	–	x? P?
Leucothoe leucothoe	–	some spp.	some spp.	x?	–	x?
Nicotiana tabacum tobacco	occ.	x	x	–	–	x? one report
Papaver somniferum opium poppy	x	–	x	–	–	x P (narcotic)
Phellodendron amurense amur cork tree	occ.	–	x	–	x?	x? D? H?
Pieris pieris	some spp.	some spp.	some spp.	x?	–	x?
Quercus oak	some spp.	some spp.	some spp.	–	–	x?[d]
Ranunculus buttercup etc.	–	some spp.	some spp.	–	x?	x P?
Rhododendron arboreum tree rhododendron	–	–	occ. B.C.	x?	–	x NP
Rhododendron lutea	–	–	–	x?	–	x NP
Rhododendron occidentalis western azalea	–	x	x	x?	–	x NP
Rhododendron ponticum Pontic rhododendron	–	–	x	x?	–	x NP

TABLE 9. Continued

Names of plants	Naturalized in North America	Native to North America	Cultivated in Canada	Honey toxic to humans	Plant poisonous to mammals	Plant poisonous to bees
Rhododendron prattii	–	–	–	x?	–	x NP
Rhododendron thomsonii	–	–	occ. B.C.	x?	–	x NP
Sapindus marginatus soapberry	–	x	–	–	–	x N
Scabiosa atropurpurea sweet scabious	x	–	x	–	–	x?
Solanum nigrum black nightshade	x	–	–	–	x	x
Sophora japonica Japanese sophora	–	–	x	–	x?	x?
Stachys arvensis field nettle	–	x	–	–	x?	x^e
Syringa amurense lilac	–	–	x	–	–	x? P?
Taxus yew	–	–	x	–	x	x P
Tilia basswood, lime	x	some spp.	some spp.	–	–	x occ.
Tulipa tulip	–	–	x	–	–	x occ.
Veratrum hellebore	–	some	–	–	x?	x
Zigadenus venenosus meadow death camus	–	x	–	–	x bulb	x

[a] frequently insects feeding on milkweed are poisonous to mammals.
[b] bees may be trapped by sticky pollinia, but spp. not poisonous.
[c] poisonous to pigmented animals only.
[d] associated fungus may be the cause of poisoning.
[e] associated yeast may be poisonous.

11. PESTICIDES AND HONEYBEES

In the most intensively farmed parts of Canada honeybees are subjected to the continual hazard of chemical poisoning that "overshadows all other problems, including bee diseases"[6].

Foraging honeybees may come into contact with pesticides deposited on plant surfaces, on soil particles, in water supplies or in the air. Most often pesticides kill only field bees thereby weakening the colony, but not seriously affecting its life. If pesticides are carried by foragers into the colony, the situation is more serious. Brood, non-foraging workers and the entire population may eventually be destroyed when pesticides are actively collected by foragers.

11.1 Pesticide toxicity to honeybees

Since many pesticides are toxic to honeybees and other beneficial insects, every attempt should be made to use only the recommended chemicals which are relatively safe for these beneficial species. Tables 11, 12, 13 and 14 list pesticides (including insecticides and acaricides) in 4 groups according to their toxicity to honeybees[a]. Each pesticide is given the standard name according to the International Standardisation Organization. If no standard common name exists, then another, non-standard common name or trade name is used.

The toxicity of any pesticide to honeybees is at least partially dependent upon environmental conditions. However, only limited research has been done on this subject. The pesticides listed below in Table 10 cannot be allocated to a toxicity group until further data is available[2].

TABLE 10. Registered pesticides of undetermined hazard to honeybees in 1986 (from Adey *et al.*[2])

alphamethrin (Fastac)	mecarbam (Murfotox)
bromophos	perchlordecone (Mirex)
bromophos-ethyl	prothiophos
dioxacarb (Famid, Gamid)	tecnazene (Fusarex)

---

[a] Tables 11 - 14 were originally compiled by Adey *et al.*[2] and are derived from information from a number of comprehensive sources. The reader is referred to this reference for further information on pesticides and honeybees.

TABLE 11. Pesticides most toxic to honeybees. For minimal hazard to honeybees, these pesticides should not be applied on blooming crops or weeds. Their residual activity is usually high, even after 10 hours (Adey et al.[2]).

D = dust
EC = emulsifiable concentrate
F = flowable
G = granular
LS = liquid suspension
MA = concentrate application at mosquito abatement rates

SP = soluble powder
ULV = ultra low volume
WP = wettable powder

Any concentration given refers to active ingredient per unit area.

---

acephate	formothion
aldicarb G	gamma-HCH
(applied 4 weeks before bloom)	heptachlor
aldrin	heptenophos
aminocarb (1 kg/ha or more)	isobenzan
azinphos-ethyl	lead arsenate
azinphos-methyl	malathion D
bendiocarb	malathion ULV (0.5 kg/ha or more)
calcium arsenate	methamidophos
carbaryl ULV (over 0.5 kg/ha)	methidathion
carbaryl D, WP, XLR (over 1.5 kg/ha)	methiocarb
carbofuran F	methomyl D
carbophenothion D	mevinphos
carbosulfan	monocrotophos
chlorpyrifos	naled D, WP
crotoxyphos	omethoate
cypermethrin (over 0.025 kg/ha)	parathion
deltamethrin	parathion-methyl
diazinon	*permethrin
dicapthon	phenthoate
dichlorvos	phosmet
dicrotophos	phosphamidon
dieldrin	phoxim
dimethoate	pirimiphos-ethyl
dinoseb	pirimiphos-methyl
DNOC (more than 0.4% dilution)	propoxur
EPN	quinalphos
etrimphos	resmethrin
fenamiphos	sulfotep
fensulfothion	tetrachlorvinphos (at higher rates)
fenthion	thionazin
fenvalerate (over 0.1 kg/ha)	triazophos
flucythrinate	vamidothion

---

* see also Table 14. More research is needed on this pesticide to establish toxicity under different conditions[2].

TABLE 12. Pesticides very toxic to honeybees.
For minimal hazard to honeybees, these pesticides should be applied only during the late evening. Their residual activity is usually low within 8 hours (Adey et al.[2]).
See Table 11 for meaning of abbreviations.

carbaryl XLR (1.5 kg/ha or less)
endosulfan (more than 0.5 kg/ha)
fenvalerate (0.1 kg/ha or less)
malathion EC
naled EC
oxamyl (1 kg/ha or more)
phorate EC
profenophos [provisional]
thiometon

TABLE 13. Pesticides less toxic to honeybees.
For minimum hazard to honeybees, these pesticides should be applied only during late evening, night or early morning. Their residual activity is usually low within 3 hours (Adey et al.[2]).
See Table 11 for meanings of abbreviations.

aminocarb ULV (0.15 kg/ha or less)
binapacryl
carbaryl ULV (0.5 kg/ha or less)
carbophenothion (not D)
chlordane
chlorfenvinphos
chlorpyrifos MA
chlorpyrifos ULV (0.05 kg/ha or less)
coumaphos
cypermethrin (0.025 kg/ha or less)
DDT
demeton
demeton-S-methyl
dichlofenthion
dichlorvos MA (0.05 kg/ha or less)
dieldrin G
dimetilan
dinobuton
dioxathion
disulfoton EC
DNOC (less than 0.4% dilution)
endosulfan (0.5 kg/ha or less)
endrin
ethiofencarb
ethion
fenchlorphos
fenthion G, MA

fonofos
formetanate
heptachlor G
leptophos
malathion MA
malonoben
menazon
methomyl LS, SP
methoxychlor
Nissol
oil sprays (superior type)
oxamyl (0.5 kg/ha or less)
oxydemeton-methyl
phorate G
phosalone
pirimicarb
propoxur MA
pyrazophos [provisional]
rotenone D
temephos
TEPP
tetrachlorvinphos (at lower rates)
thiodicarb
toxaphene
trichlorfon

TABLE 14. Pesticides least toxic to honeybees.
These pesticides can be applied at any time with reasonable safety to honeybees. Their toxicity is usually low with direct application (Adey *et al.*[2]).
See Table 11 for meanings of abbreviations.

allethrin	fenazaflor
amitraz	fenbutatin oxide
azocyclotin	fenoxycarb
Bacillus thuringiensis	fenson
bromopropylate	fensulfothion G
carbaryl G	fluvalinate
carbofuran G	*Heliothis* virus
chinomethionate	hydroprene
chlofentezin	isofenphos
chlorbenside	lime sulphur
chlordecone	malathion G
chlordimeform	mancozeb
chlorfenethol	mirex G
chlorfenson	nicotine sulphate
chlorfensulphide	*permethrin
chlorobenzilate	propargite
chlorpropylate	propoxur G
cryolite	pyrethrum
cyhexatin	rotenone EC
dicofol	ryania
dienochlor	schradan
diflubenzuron	sodium fluorosilicate baits
dinocap	sulphur
disulfoton G	tetradifon
fenazaflor	thiocyclam

* Reported to be made safe by its repellency under arid conditions; see Table 10.1.

11.2 Precautions to avoid serious loss

A gardener, farmer, or beekeeper may occasionally need to apply a pesticide known to be harmful to bees. In order to avoid serious loss of foragers and colonies in the vicinity, the following precautions are suggested[6]. Although these precautions are listed in an approximate order of importance, the particular circumstances will determine which combination of measures would be appropriate.

1. Beekeepers should be notified before any pesticide hazardous to bees is applied to crops. Beekeepers must make any person applying pesticides aware of the need for such notification.

2. Bees can be moved away from the area. Usually injury is not significant to colonies placed a quarter of a mile (0.4 km) or more away from treatments, unless the treated crop is the only attractive source in the area. In this case, colonies need to be moved further than several kilometres away.

3. Bees should be moved away from crops being pollinated when 90% petal fall occurs, or as soon as possible when pollination is accomplished. It is important to remove bees as quickly as possible from any potentially hazardous situation.

4. Treatments made during the night are safest, or during the early morning before bees are foraging.

5. Colonies can be covered with burlap or dark plastic material for 1 to 2 hours, during and after treatment in early mornings to give added protection. Colonies may be kept covered for up to 2 days if the covering is kept moist.

6. Non-blooming crops should not be treated while cover crops, weeds or wild flowers are blooming in a field nearby. These adjacent blooms should be removed first if spraying the crop is necessary.

7. Crops which are attractive to bees should never be sprayed while blooming.

8. Pesticide drift should be reduced. Pesticides should not be applied when there is a significant breeze. This is especially important when applying formulations of fine sprays and dusts by aircraft.

In general, some preparations and formulations are more hazardous than others to honeybees. Those which are potentially less harmful should always be used on crops, garden plants or weeds visited by bees. Granular formulations (except micro-encapsulates) are usually safer than any other. Table 15 differentiates between the more hazardous and less hazardous formulations.

TABLE 15. Formulations and preparations of pesticides listed according to their potential hazard to honeybees (adapted from Atkins[6]).

Usually more hazardous	Usually less hazardous
inorganic arsenicals	organic pesticides
combination of toxic pesticides	single toxic pesticide
preparations of dusts	preparations of sprays
wettable powder formulations	emulsifiable or water soluble concentrates
coarse sprays	fine sprays
spray applications by aircraft over bees in flight	application by ground equipment
micro-encapsulates	granular applications

12. CANADIAN BEEKEEPING

> "....Gitche Manito, the Mighty
> The Great Spirit, The Creator,
> Sends them hither on his errand,
> Sends them to us with his message.
> Whereso'er they move, before them
> Swarms the stinging fly, the Ahmo,
> Swarms the bee, the honey-maker;
> Springs a flower unknown among us,
> Springs the White-Man's foot in blossom."
> Longfellow, H.W., *Hiawatha*

Canada is a country of modern commercial beekeeping, and has no indigenous or traditional beekeeping background such as there is in Europe, Africa or Asia. In fact, as Longfellow describes so poetically in this excerpt from *Hiawatha*, honeybees were unknown to the indigenous population until introduced by the early settlers, perhaps 250 years ago. These so called "stinging flies" adapted easily to the New World environment, and are now successfully established in feral colonies and apiaries over both the Americas from Argentina to Alaska.

12.1 Canada's position as a world honey producer

Today,[a] Canada is one of the most progressive beekeeping countries of the world. On the basis of honey yield per colony Canada consistently has the highest national average (see Table 16). In terms of total production Canada is a major world producer of honey, ranking about fifth, following the People's Republic of China, the Soviet Union, the United States of America, and Mexico. About 15,000 metric tons[b] are exported annually, the majority to the United States and Japan.

Canada, like most other "developed" countries of the New World, outstrips honey production per beekeeper in the Old World by a factor of nearly six. Many of the "less developed" countries of the New World, for instance in South and Central America, and countries in Africa and Asia, still have unexploited land. When mechanization can be applied to beekeeping in these countries their *per capita* production could rise steeply. Production in Canada and the USA is also likely to increase as management practices improve, but less sharply. However, it is inevitable that this increasing production in the "developed" New World will gradually taper off, limited by a shortage of economically accessible bee forage.

It is interesting to note that although the USA and Canada cover roughly the same land area, the USA has 40 times as many beekeepers but only 10 times the colony density. Each beekeeper in the USA on average has fewer colonies than those in Canada.

---

[a] Most of the information in sections 12.1 to 12.4 comes from the following references: 12, 13, 15, 39.

[b] a metric ton is 2,204.6 lbs.

TABLE 16. A survey of world honey production. Honey production, colony numbers, yield and colony density for selected countries in 1985.

Country	Area (1000 km²)[1*]	Total Colonies (1000's)[2*]	Honey Production (metric tons)[2*]	Yield per Colony	Colony Density (colonies/km²)[1*,2*]
Argentina	2,809	1,500	30,000	26.7	5.30
Australia	7,687	525	24,490	46.6	0.07
Brazil	8,512	1,600	28,000	17.5	0.18
Canada	9,976	696	40,550	58.3	0.07
China	9,597	6,800	140,000	24.1	0.71
Germany (Federal Rep.)	249	1,150	18,000	15.7	4.62
Japan	370	285	6,200	21.8	0.77
Mexico	1,967	2,500	55,000	22.0	1.27
U.S.A.	9,660	4,325	75,000	17.3	0.44
U.S.S.R	22,400	8,150	210,00	25.8	0.36

1* Statesman's Yearbook (Paxton, 1970).
2* USDA. (1985) Sugar, molasses and honey. Foreign Agriculture
   Horticultural and Tropical Products Division, Washington, D.C.

12.2 Geology and beekeeping in Canada

Innumerable factors, including geography, climate, native flora, agricultural practices and colony management contribute to Canada's high yields of honey. Of these, the geographic factors, notably bedrock geology, are of paramount importance in determining and limiting the area available for beekeeping in Canada.

12.2.1 Pre-Cambrian Shield

A predominant feature of Canadian geology is the Pre-Cambrian Shield which covers almost half the country (Fig. 1). This feature has a tremendous influence on the nature of the landscape. The area is underlain by an ancient, hard Pre-Cambrian bedrock and is characterized by thin overburden, acid soil, muskeg, scrub vegetation and poor drainage. Not surprisingly, most of this land is not amenable to agriculture. Nevertheless, beekeeping and honey production are limited over a large part of the Shield Region largely due to the inaccessible nature of the territory, and not primarily because of lack of bee forage.

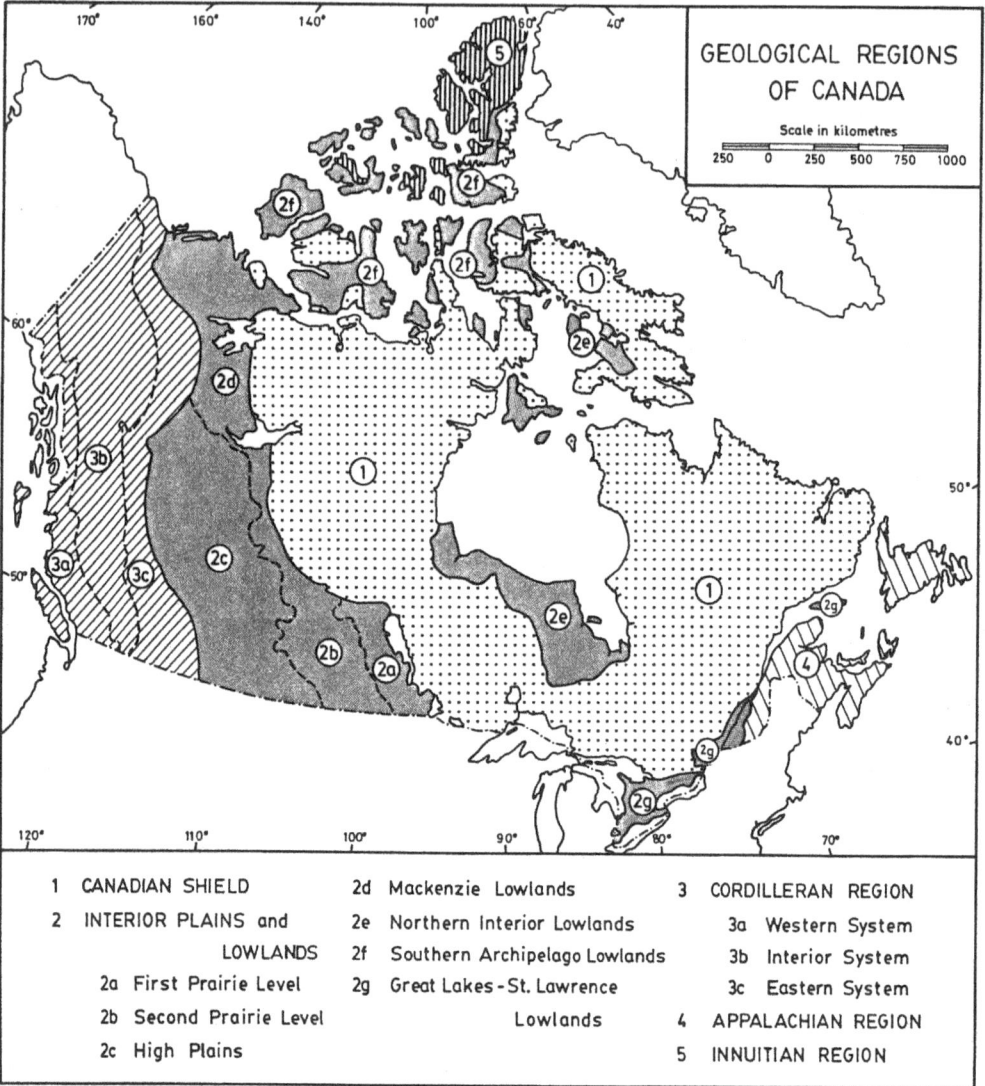

GEOLOGICAL REGIONS OF CANADA

Scale in kilometres

250 0 250 500 750 1000

1	CANADIAN SHIELD	2d	Mackenzie Lowlands	3	CORDILLERAN REGION
2	INTERIOR PLAINS and	2e	Northern Interior Lowlands		3a Western System
	LOWLANDS	2f	Southern Archipelago Lowlands		3b Interior System
	2a First Prairie Level	2g	Great Lakes - St. Lawrence		3c Eastern System
	2b Second Prairie Level		Lowlands	4	APPALACHIAN REGION
	2c High Plains			5	INNUITIAN REGION

FIG. 1. Geological regions of Canada (from Canada Year Book 1956, in Crane 1966[13]).

12.2.2 Prairie levels

Geology also defines the most productive beekeeping region, found in the Interior Plains and Lowlands (Fig. 1). A wide band of soft rock within this area, flanking the Canadian Shield to the west and south west, is known as the Prairie Levels. Some of the best agricultural soils are here, notably in the Wheat Belt running across the southern part of the Prairies. A large part of the best honey-producing country is located between the Wheat Belt and the Shield Region, especially within a 100 miles or so (161 km) of the Shield in the area known as the First Prairie Level. Specifically, the northern boundary of this area lies inside the northern edge of the boreal (mixed woodland) forest and follows the edge of the Shield. This "Honey Belt" region used to support indigenous boreal forest, but is now largely cleared for agriculture.

12.2.3 The Peace River Region

The northern limit of the Honey Belt is the Peace River country, well known for high honey yields and prosperous beekeeping. The Peace River Region is a broad fertile valley roughly 250 km across and slightly longer than 250 km, running from Fort St. John and Dawson Creek in B.C. to High Prairie in Alberta.

It is one of the most recently cleared and settled regions of Canada, and has attracted many large-scale commercial honey-producers. Fortunately for beekeepers most of the soils are more suitable for legumes and canola (rape) than wheat. At this high latitude, summer days are longer and insolation values are much higher than further south. These conditions favour very active photosynthesis, rapid plant growth and prolific nectar production. Consequently honeybee colonies prosper, and average yields are usually among the highest in the country (Table 17).

12.2.4 Cordilleran Region

Another geologic feature which exerts a definite constraint on beekeeping in Canada is the Cordilleran Region (Fig. 1), the mountain ranges which run roughly parallel to the Pacific coast. Here, both beekeeping and arable agriculture are restricted by relief and extremes of rainfall. Near the coast average annual precipitation is over 2500 mm, and in the rain shadow east of the ranges it falls below 250 mm.

TABLE 17. Canadian Beekeeping Statistics. Estimates of the number of beekeepers and colonies of bees, production and value of honey and wax in Canada[a], by Province, 1985 and 1986 with five-year averages, 1980-1984 (Statistics Canada, November 1986. Report on Honey Production and Value for 1985 and 1986. Preliminary estimate).
P = preliminary estimates

Province and Year	Colonies	Honey: Average yield per colony[b]	
		kg	(lbs.)
Prince Edward Island			
Average 1980-1984	1,330	43	(95)
1985	900	43	(95)
1986P	875	36	(80)
Nova Scotia			
Average 1980-1984	7,040	34	(75)
1985	8,000	29	(65)
1986P	6,500	21	(46)
New Brunswick			
Average 1980-1984	4,400	37	(82)
1985	4,200	36	(80)
1986P	5,000	50	(23)
Québec			
Average 1980-1984	111,600	37	(82)
1985	110,000	44	(96)
1986P	100,000	30	(66)
Ontario			
Average 1980-1984	110,400	31	(69)
1985	111,000	37	(82)
1986P	115,000	25	(55)
Manitoba			
Average 1980-1984	107,400	70	(154)
1985	120,000	73	(160)
1986P	110,000	73	(160)
Saskatchewan			
Average 1980-1984	91,200	75	(165)
1985	105,000	70	(155)
1986P	120,000	54	(120)
Alberta			
Average 1980-1984	171,200	62	(136)
1985	181,000	46	(102)
1986P	190,000	54	(120)
British Columbia			
Average 1980-1984	54,280	43	(94)
1985	53,500	43	(94)
1986P	55,000	36	(80)

TABLE 17. Continued

Province and Year	Colonies	Honey: Average yield per colony[b]	
		kg	(lbs.)
Canada			
Average 1980-1984	658,850	54	(118)
1985	693,600	52	(115)
1986P	702,375	47	(104)

[a] Does not include Newfoundland.
[b] Figures based on the commercial beekeepers' survey.
Note: 1 pound = 0.453 kilogram; 2205 pounds = 1 metric tonne.

12.3 The area available for beekeeping

Geological and geographical factors influence the total percentage of Canada's land that is available for beekeeping. Approximately 45% is uncultivated and inaccessible (mostly Shield). A further 42% is covered with boreal forest (predominantly coniferous) and 6% with mixed hardwood (deciduous) trees. The dense coniferous forests yield little or no floral nectar except in areas of tree felling, where pioneer species such as fireweed (*Epilobium angustifolium*) may offer lucrative short-term rewards. The potential of honeydew production from these areas is largely unexplored and at present unexploited (see Section 9).

The remaining 7% of Canada is agriculturally productive. Since approximately one quarter of this area is devoted to grain (notably wheat), this leaves a very small total area with potential for bee forage. As a result colony density (number of colonies/total area) for the country as a whole is lower than any other except Australia (see Table 16).

12.4 Major beekeeping areas and honey producing plants

The bulk of surplus in Canada comes from agricultural crops. The following are important nectar sources and are grown in all provinces.

alsike clover	(*Trifolium hybridum*)
bird's foot trefoil	(*Lotus corniculatus*)
Ladino clover	(*Trifolium repens* Forma *ladigense*)
red clover	(*Trifolium pratense*)
sweet clover	(*Melilotus* spp.)

Other cultivated crops, important only in certain regions, include:

buckwheat	(*Fagopyrum esculentum*)
canola and mustard	(*Brassica* spp.)
sunflower	(*Helianthus* spp.)

The active season for foraging commences in late April in all provinces in Canada, except in coastal B.C. where the season is usually a few weeks earlier. The main flow begins during the third week of June in most areas, but is a few weeks later in the most northerly beekeeping regions.

For the sake of convenience Canada may be divided into a number of major honey producing areas. Table 18 describes 10 honey producing regions and indicates the approximate phenology (flowering period) of the major honey plants in each region.

TABLE 18. Approximate phenology of the ten most important nectar or pollen sources in the major honey producing areas of Canada (adapted from unpublished table by Colter, D. 1976). N nectar, P pollen, x in flower, * offering N or P

Plant Names Area:	Value		Months											
: Vancouver Island[a]	P	N	J	F	M	A	M	J	J	A	S	O	N	D
willow (*Salix* spp.)	*	*			x	x	x							
broom (*Cytisus*)	*					x	x	x						
big-leaf maple (*Acer*)	*	*			x	x								
dandelion (*Taraxacum* spp.)	*	*				x	x							
arbutus (*Arbutus*)	*	*					x							
snowberry (*Symphoricarpos*)		*					x	x						
blackberry (*Rubus* spp.)	*	*						x	x					
salal (*Gaultheria*)		*						x	x					
thistle (*Cirsium* spp.)	*	*						x	x	x				
fireweed (*Epilobium*)	*	*						x	x	x				
: Okanagan Valley[b]	P	N	J	F	M	A	M	J	J	A	S	O	N	D
willow (*Salix* spp)	*	*				x	x							
dandelion (*Taraxacum* spp.)	*	*				x	x							
fruit trees (*Prunus, Malus*)	*	*				x	x							
snowberry (*Symphoricarpus*)		*					x	x	x					
sweet clover (*Melilotus*)	*	*						x	x	x				
alfalfa (*Medicago*)		*						x	x	x				
wild mustard (*Brassica*)	*	*						x	x	x				
knapweed (*Centaurea*)	*	*							x	x				
fireweed (*Epilobium*)	*	*								x	x	x		
: Peace River District[b]	P	N	J	F	M	A	M	J	J	A	S	O	N	D
willow (*Salix* spp.)	*	*				x	x							
dandelion (*Taraxacum*)	*	*				x	x							
silverberry (*Elaeagnus*)	*	*						x						
sweet clover (*Melilotus*)	*	*						x	x	x				
alfalfa (*Medicago*)		*							x	x				
clover (*Trifolium* spp.)	*	*						x	x	x				
canola (*Brassica*)	*	*						x	x	x				
fireweed (*Epilobium*)	*	*							x	x				
thistle (*Cirsium* spp.)	*	*							x	x				
sowthistle (*Sonchus*)	*	*							x	x				

TABLE 18. Continued

:Southern Alberta[c]

	P	N	J	F	M	A	M	J	J	A	S	O	N	D
willow (*Salix* spp.)	*	*			x	x	x							
dandelion (*Taraxacum*)	*	*				x	x							
serviceberry (*Amelanchier*)	*	*					x	x						
canola (*Brassica*)	*	*					x	x						
sainfoin (*Onobrychis*)	*	*						x	x					
sunflower (*Helianthus* spp.)	*	*						x	x					
clover (*Trifolium* spp.)	*	*						x	x					
sweet clover (*Melilotus*)	*	*						x	x	x				
alfalfa (*Medicago*)		*						x	x	x				

: Central Alberta[c]

	P	N	J	F	M	A	M	J	J	A	S	O	N	D
willow (*Salix* spp.)	*	*				x	x							
poplar (*Populus* spp.)	*					x	x							
dandelion (*Taraxacum*)	*	*					x							
saskatoon (*Amelanchier*)	*	*					x	x						
canola (*Brassica*)	*	*					x	x						
sweet clover (*Melilotus*)	*	*						x	x	x				
clover (*Trifolium* spp.)	*	*						x	x	x				
goldenrod (*Solidago* spp.)	*	*								x	x			
aster (*Aster* spp.)	*	*								x	x			

: Northern Alberta - Peace River District[c]

	P	N	J	F	M	A	M	J	J	A	S	O	N	D
willow (*Salix* spp.)	*	*				x	x							
poplar (*Populus* spp.)	*					x	x							
dandelion (*Taraxacum*)	*	*					x							
canola (*Brassica*)	*	*					x	x						
sweet clover (*Melilotus*)	*	*						x	x	x				
clover (*Trifolium* spp.)	*	*						x	x	x				
hawk's beard (*Crepis*)	*	*						x	x					
alfalfa (*Medicago*)		*							x	x				
fireweed (*Epilobium*)	*	*							x	x	x			
aster (*Aster* spp.)	*	*							x	x				

: North-Central Saskatchewan[d]

	P	N	J	F	M	A	M	J	J	A	S	O	N	D
willow (*Salix* spp.)	*	*					x							
poplar (*Populus* spp.)	*						x							
dandelion (*Taraxacum*)	*	*					x	x	x					
Saskatoonberry (*Amelanchier*)	*	*					x	x						
canola (*Brassica*)	*	*					x	x						
clover (*Trifolium* spp.)	*	*						x	x	x				
alfalfa (*Medicago*)		*						x	x	x				
giant hyssop (*Agastache*)	*	*							x	x				
fireweed (*Epilobium*)	*	*							x	x				
aster (*Aster* spp.)	*	*							x	x	x			

TABLE 18. Continued

: South-East Saskatchewan- South-West Manitoba[d]	P	N	J	F	M	A	M	J	J	A	S	O	N	D
willow (*Salix* spp.)	*	*					x	x						
dandelion (*Taraxacum*)	*	*					x	x						
caragana (*Caragana*)	*	*					x	x						
snowberry (*Symphoricarpos*)	*	*					x	x						
canola (*Brassica*)	*	*					x	x						
clover (*Trifolium* spp.)	*	*						x	x					
sweet clover (*Melilotus*)	*	*						x	x					
sunflower (*Helianthus*)	*	*						x	x					
alfalfa (*Medicago*)		*						x	x	x				
goldenrod (*Solidago*)	*	*							x	x	x			

: Central Manitoba[c]	P	N	J	F	M	A	M	J	J	A	S	O	N	D
willow (*Salix* spp.)	*	*					x	x						
poplar (*Populus* spp.)	*						x	x						
dandelion (*Taraxacum*)	*	*					x	x						
buckwheat (*Fagopyrum*)		*						x	x					
sweet clover (*Melilotus*)	*	*						x	x	x				
clover (*Trifolium* spp.)	*	*						x	x	x				
giant hyssop (*Agastache*)		*						x	x	x				
thistle (*Cirsium* spp.)	*	*						x	x	x				
goldenrod (*Solidago*)	*	*						x	x	x				
aster (*Aster* spp.)	*	*							x	x	x			

: St. John River Valley[e]	P	N	J	F	M	A	M	J	J	A	S	O	N	D
willow (*Salix* spp.)	*	*					x	x						
deciduous trees (*Acer*, *Populus*)	*	*					x	x						
fruit trees (*Malus*, *Prunus*)	*	*					x	x						
blueberry (*Vaccinium*)		*					x	x						
clover (*Trifolium*, *Melilotus*)	*	*						x	x					
buckwheat (*Fagopyrum*)		*						x	x					
thistles (*Cirsium*)	*	*							x	x				
fireweed (*Epilobium*)	*	*							x	x	x			
goldenrod (*Solidago*)	*	*								x	x	x		
aster (*Aster* spp.)	*	*								x	x	x		

: Annapolis Valley[f]	P	N	J	F	M	A	M	J	J	A	S	O	N	D
willow (*Salix* spp.)	*	*					x	x						
fruit tree (*Malus*, *Prunus* spp.)	*	*					x	x						
dandelion (*Taraxacum*)	*	*					x							
sumac (*Rhus*)		*					x	x						
clover (*Trifolium*, *Melilotus*)	*	*						x	x					
berry fruit (*Rubus*, *Vaccinium*)	*	*						x	x					
sow thistle (*Cirsium*)	*	*						x	x					
goldenrod (*Solidago* spp.)	*	*							x	x	x			
aster (*Aster* spp.)	*	*								x	x	x		

TABLE 18. Continued

: Prince Edward Island[g]

	P	N	J	F	M	A	M	J	J	A	S	O	N	D
willow (*Salix* spp.)	*	*				x	x							
deciduous trees (*Alnus, Populus*)	*					x	x							
maple (*Acer* spp.)	*	*				x	x							
elderberry (*Sambucus*)	*	*				x	x	x						
dandelion (*Taraxacum*)	*	*				x	x	x						
blueberry (*Vaccinium*)	*	*					x	x						
clover (*Trifolium* spp.)	*	*					x	x	x					
fireweed (*Epilobium*)	*	*						x	x	x				
goldenrod (*Solidago*)	*	*								x	x	x		
aster (*Aster* spp.)	*	*									x	x		

: South-Western Ontario[c]

	P	N	J	F	M	A	M	J	J	A	S	O	N	D
willow (*Salix* spp.)	*	*			x	x	x							
deciduous trees (*Acer, Robinia*)		*				x	x	x						
dandelion (*Taraxacum*)	*	*				x	x							
canola (*Brassica*)	*	*					x	x						
white clover (*Trifolium*)	*	*						x	x					
bird's foot trefoil (*Lotus*)		*						x	x					
alfalfa (*Medicago*)		*						x	x	x				
purple loosestrife (*Lythrum*)	*	*						x	x					
milkweed (*Asclepias*)	*	*						x	x					
goldenrod (*Solidago*)	*	*								x	x	x		

: Niagara Peninsula[c]

	P	N	J	F	M	A	M	J	J	A	S	O	N	D
willow (*Salix* spp.)	*	*			x	x	x							
deciduous trees (*Acer, Sambucus*)	*	*				x	x							
dandelion (*Taraxacum*)	*	*				x	x							
fruit trees (*Malus, Prunus*)		*				x	x							
berry fruits (*Rubus*)		*					x	x						
clover (*Trifolium* spp.)	*	*						x	x					
alfalfa (*Medicago*)		*						x	x					
basswood (*Tilia*)		*						x	x					
goldenrod (*Solidago*)	*	*								x	x	x		
aster (*Aster* spp.)	*	*									x	x	x	

: Ottawa Valley[c]

	P	N	J	F	M	A	M	J	J	A	S	O	N	D
willow (*Salix* spp.)	*	*				x	x							
fruit trees (*Malus, Prunus*)	*	*				x	x							
dandelion (*Taraxacum*)	*	*					x							
hawthorn (*Crataegeus*)	*	*					x	x						
blueweed (*Echium*)		*						x	x					
clover (*Trifolium* spp.)	*	*						x	x	x				
sweet clover (*Melilotus*)	*	*						x	x	x				
fireweed (*Epilobium*)	*	*							x	x				
goldenrod (*Solidago*)	*	*								x	x	x		
aster (*Aster* spp.)	*	*									x	x	x	

TABLE 18. Continued

: Upper St. Lawrence- Gaspé Peninsula[h]	P	N	J	F	M	A	M	J	J	A	S	O	N	D
willow (*Salix* spp.)	*	*		x	x	x								
dandelion (*Taraxacum*)	*	*				x	x	x	x	x				
maple (*Acer*)	*	*				x	x							
fruit trees (*Malus, Prunus*)	*	*				x	x							
raspberry (*Rubus*)	*	*					x	x	x					
alfalfa (*Medicago*)		*					x	x	x	x				
clover (*Trifolium*)		*					x	x	x					
sweet clover (*Melilotus*)		*					x	x	x					
goldenrod (*Solidago*)	*	*								x	x			
aster (*Aster*)	*	*								x	x			

[a] Reviewed by Stephen Mitchell
[b] Reviewed by Douglas McCutcheon and Douglas Colter
[c] Reviewed by Douglas Colter
[d] Reviewed by John Gruszka and Douglas Colter
[e] Reviewed by Bruce Palmer
[f] Reviewed by Lorne Crozier
[g] Reviewed by Larry Cosgrave
[h] Reviewed by Jean-Louis Villeneuve

13. USEFUL LISTS

13.1 Early Pollen Sources

a) bulbs:

Chionodoxa luciliae	glory of the snow
Crocus	crocus
Galanthus nivalis	snow-drop
Leucojum vernum	spring snowflake
Muscari botryoides	grape hyacinth
Scilla siberica	Siberian squill

b) herbaceous perennials:

Anemone nemorosa	wood anemone
Anemone patens	Prairie crocus
Arabis caucasica	wall rock cress
Eranthis hyemalis	winter aconite

c) trees and shrubs:

Alnus	alder
Amelanchier alnifolia	Saskatoon berry
Berberis darwinii	Darwin barberry
Berberis × stenophylla	rosemary barberry
Corylus	hazelnut, filbert
Daphne mezereum	February daphne
Diervilla lonicera	honeysuckle diervilla
Erica carnea	spring heath
Hamamelis mollis	Chinese witch hazel
Lonicera involucrata	twin berry
Mahonia aquifolium	Oregon grape
Populus	poplar
Prunus	cherry, plum etc
Pyrus	pear
Quercus	oak
Salix	willow
Shepherdia argentea	silver-leaved buffalo berry
Ulmus americana	American elm
Viburnum tinus	laurestinus

13.2 Some of the Most Attractive Nectar Sources

These plants have excellent reputations as nectar sources. Most of the species are suitable for amenity planting in private and public gardens.

* denotes an outstanding reputation as a bee plant (chiefly because of its nectar rather than its pollen).

a) annuals and biennials:

Borago officinalis	borage
Brassica napus spp. *napus*	canola
Brassica rapa spp. *oleifera*	canola
Carthamnus tinctoria	safflower
Cleome hasslerana	spider plant, spider flower
Cleome serrulata	Rocky Mountain bee plant
Coriandrum sativum	coriander
Digitalis	foxglove
Dispsacus sativus	teasel
Dracocephalum moldavica	annual dragon's head
Echium vulgare	viper's bugloss
Fagopyrum	buckwheat
Gaillardia pulchella	painted gaillardia
Helianthus petiolaris	Prairies sunflower
Leonurus sibericus	chivirico
Melilotus	sweet clover
Origanum majorana	sweet marjoram
Phacelia tanacetifolia	phacelia
Pimpinella anisatum	anise
Reseda odorata	mignonette
Satureia hortensis	summer savory
Trifolium incarnatum	crimson clover
Vicia pannonica	Hungarian vetch
Vicia sativa	common vetch
Vicia villosa	hairy vetch

b) perennials:

Agastache foeniculum	anise hyssop
Ajuga	bugle
Allium	onion
Aster	aster
Aubretia deltoidea	purple rock cress
Baptisia tinctoria	wild indigo
Centaurea	centaurea, bluet
Echinops sphaerocephalus	globe thistle
Epilobium angustifolium	fireweed
Hedera helix	English ivy
Hyssopus officinalis	garden hyssop
Leonurus cardiaca	common motherwort
Lythrum salicaria	purple loosestrife
Marrubium vulgare	white horehound
Medicago sativa	alfalfa
Mentha	mint
Monarda punctata	horse mint

*Nepeta cataria	catnip
Onobrychis viciifolia	sainfoin
Origanum vulgare	wild marjoram
Pycnanthemum virginianum	mountain mint
Salvia	sage
*Scrophularia marilandica	Maryland figwort

c) trees and shrubs:

*Acer	maple
Aesculus	chestnut
*Ailanthus altissima	tree of heaven
Aralia	angelica tree
Caragana aborescens	Siberian peashrub, caragana
Cotoneaster	cotoneaster
Elaeagnus angustifolia	Russian olive
Evodea daniellii	Korean evodea
*Ilex	holly
*Koelreuteria paniculata	golden rain tree
Lavandula	lavender
*Liriodendron tulipifera	tulip tree
Malus	apple, crab apple
*Paulownia tomentosa	princess tree
Perovskia atriplicifolia	silver sage
*Ptelea trifolicata	hop tree
*Robinia	acacia
Rosmarinus officinalis	rosemary
Rubus	blackberry, raspberry etc
Salix	willow
Sophora japonica	Japanese pagoda tree
Symphoricarpos albus	snowberry
*Tilia	basswood, lime, linden
Vaccinium corymbosum	high bush blueberry
Vitex negundo cv. heterophylla	cut-leaved chaste tree

13.3 A Selection of Plants with Potential for Roadsides and Banks

** denotes plants especially useful for erosion control

**Ajuga reptans	bugleweed
**Cornus stolonifera	red osier dogwood
**Diervilla lonicera	honeysuckle diervilla
**Diervilla sessifolia	southern bush honeysuckle
**Gleditsia triacanthos	common honey locust
Hedera helix	English ivy
Hedysarum coronarium	sweet vetch
**Lathyrus tuberosus	ground nut pea
**Lathyrus sylvestris 'Wagneri'	Wagner pea
**Lespedeza bicolor	shrub bush clover
Lotus corniculatus	bird's foot trefoil
Lonicera morrowii	Morrow honeysuckle
Lupulina angustifolia	lupine (field crop)

Medicago falcata	Siberian yellow flowered alfalfa
***Medicago lupulina*	black medick
Medicago sativa	alfalfa
Melilotus	sweet clover
Parthenocissus quinquefolia	Virginia creeper
***Polygonum cuspidatum*	Japanese knotweed
Rhus glabra	smooth sumac
**Robinia hispida	rose acacia
***Robinia pseudoacacia*	black locust
**Rosa multiflora	Japanese rose, multiflora rose
***Symphoricarpos orbiculatus*	coral berry
Trifolium spp.	true clover
Vicia pannonica	Hungarian vetch
Vicia villosa	hairy vetch

13.4 Calendar of plants

Perhaps the single most important piece of information about any bee forage plant, after its potential to increase the honey surplus, is when it will start blooming. The lists in Sections 13.4.1, 13.4.2 and 13.4.3 give a rough idea of plant blooming times and can be consulted before investigating a plant in detail. Plants are grouped according to the season when they begin blooming, (very early, early, moderately early, mid, late, and very late). The month which is approximately equivalent to the start of a particular season is indicated in each list.

Actual dates of first bloom vary depending on local growing conditions (see Section 4), and are different from year to year in any one location. However, plants that usually bloom together will do so regardless of where they are growing. A few plants, especially among the annuals and biennials, have very long blooming periods (i.e. 2 or more months) and are "good value" for gardeners and beekeepers. Trees and shrubs tend to have short blooming periods (i.e. 1-2 weeks). Perennials vary enormously, but usually flower for longer than most trees or shrubs and for shorter than most annuals or biennials.

13.4.1 Calendar of annuals and biennials

Approximate season of first bloom of most annuals and biennials listed in the index:

EARLY (May)
Campanula medium	Canterbury bell
Cheiranthus cheiri	true wall-flower
Cleome lutea	yellow spider flower
Nemophila menziesii	baby blue eyes
Trifolium incarnatum	crimson clover

MID (June)
Alcea rosea	hollyhock
Borago officinalis	borage
Brassica napus ssp. *napus*	canola
Brassica rapa ssp. *oleifera*	canola

Centaurea cyanus	cornflower
Cheiranthus allioni	Siberian wall-flower
Clarkia elegans	rose clarkia
Coriandrum sativum	coriander
Digitalis	foxglove
Dipsacus sativus	fuller's teasel
Echium vulgare	viper's bugloss
Gaillardia pulchella	painted gaillardia
Isatis tinctoria	dyer's woad
Limnanthes douglasii	meadow foam
Linum usitatissimum	flax
Lobualaria maritimum	alyssum
Melilotus	sweet clover
Nicotiana alata	winged tobacco
Papaver rhoeas	corn poppy
Reseda odorata	mignonette
Scabiosa atropurpurea	sweet scabious
Senecio elegans	purple ragwort
Verbascum blatteria	moth mullein
Vicia faba	broad bean
Vicia pannonica	Hungarian vetch
Vicia sativa	common vetch
Vicia villosa	hairy vetch
LATE (July)	
Alcea ficifolia	figleaf hollyhock
Brassica nigra	black mustard
Cleome serrulata	bee spider flower
Centaurea moschata	sweet sultan
Convolvulus tricolor	morning glory
Coreopsis	tickseed
Dracocephalum moldavica	annual dragon's head
Eschscholzia californica	California poppy
Fagopyrum esculentum	buckwheat
Fagopyrum tataricum	buckwheat
Gilia cap itata	globe gilia
Helianthus annuus	common sunflower
Helianthus petiolaris	prairie sunflower
Iberis umbellata	globe candy tuft
Lavatera trimestris	herb tree mallow
Origanum majorana	sweet sultan
Phaseolus coccineus	scarlet runner bean
Salvia sclarea	clary sage
Satureja hortensis	summer savory
Sinapis alba	white mustard
Tagetes erecta	African marigold
Zinnia elegans	common zinnia
VERY LATE (August or September)	
Cleome hasslerana	spiny spider flower
Cosmos	cosmos
Impatiens glandulifera	Himalayan balsam
Leonurus sibericus	chivirico
Limonium sinuatum	statice
Melilotus alba	white sweet clover
Melilotus alba 'Annua'	Hubam clover
Nicotiana tabacum	tobacco

Ocimum basilicum	sweet basil
Pimpinella anisum	anise
Rudbeckia hirta	black eyed Susan

13.4.2 Calendar of perennials

Approximate season of first bloom of most perennials listed in the index:

VERY EARLY (March)

Arabis caucasica	wall rock cress
Eranthis hyemalis	winter aconite
Galanthus nivalis	snowdrop

EARLY (April)

Anemone nemorosa	European wood anemone
Anemone patens	prairie crocus
Chionodoxa luciliae	glory of the snow
Claytonia virginica	Virginia spring beauty
Crocus	crocus
Hyacinthus orientale	Oriental hyacinth
Leucojum vernum	spring snowflake
Muscari botryoides	grape hyacinth
Scilla siberica	Siberian squill

MODERATELY EARLY (May)

Ajuga genevensis	Geneva bugle
Ajuga reptans	bugle weed
Alyssum saxatile	basket of gold
Anemone blanda	Greek anemone
Armeria maritima	common thrift
Aubretia deltoidea	purple rock cress
Baptisia australis	blue wild indigo
Camassia esculenta	quamash
Clematis armandii	Armand clematis
Doronicum plantagineum	leopard's bane
Geranium pratens	meadow crane's bill
Geum	avens
Medicago sativa	alfalfa
Papaver orientale	Oriental poppy

MID (June)

Agastache integrifolia	anise hyssop
Allium giganteum	giant onion
Anchusa azurea	Italian bugloss
Campanula carpatica	Carpathian bellflower
Centaurea dealbata	Persian centaurea
Centaurea montana	mountain bluet
Dictamnus albus	burning bush
Geranium ibericum	Iberian geranium
Hedysarum coronarium	French honeysuckle
Hydrophyllum virginianum	Virginia water leaf
Hyssopus officinalis	hyssop
Lathyrus tuberosus	ground nut pea
Leonurus cardiaca	motherwort
Linum flavum	golden flax
Linum perenne	perennial flax

Lotus corniculatus	bird's foot trefoil
Lysimachia punctata	yellow loosestrife
Lythrum salicaria	purple loosestrife
Marrubium vulgare	white horehound
Nepeta mussinii	Persian catnip
Oenothera	evening primrose
Onobrychis viciifolia	sainfoin
Opuntia polycantha	prickly pear cactus
Paeonia	peony
Penstemon barbatus	bearlip penstemon
Penstemon grandiflorus	large flowered penstemon
Polemonium caeruleum	Jacob's ladder
Ranunculus	buttercup
Salvia officinalis	garden sage
Salvia pratensis	meadow sage
Salvia × superba	salvia
Thalictrum	rue
Thymus serpyllum	wild thyme
Thymus vulgaris	garden thyme
Trifolium hybridum	alsike clover
Trifolium pratens	red clover
Trillium	trillium

MODERATELY LATE (July)

Anaphalus margaritacea	pearly everlasting
Baptisia tinctoria	wild indigo
Celastrus scandens	American bittersweet
Cichorium intybus	common chicory
Cirsium	thistle
Clematis montana	anemone clematis
Dahlia	dahlia
Echinops exaltatus	Russian globe thistle
Echinops ritro	small globe thistle
Echinops sphaerocephalus	globe thistle
Epilobium angustifolium	fireweed
Erigeron speciosus	fleabane
Erygnum maritimum	sea holly
Eupatorium perfoliatum	boneset
Foeniculum vulgare	fennel
Galtonia candicans	giant summer hyacinth
Gypsophila paniculata	baby's breath
Lavatera thuringiaca	Siberian rose mallow
Liatris pychnostachya	Kansas gayfeather
Limonium latifolium	statice latifolium
Malva moschata	musk mallow
Mentha arvensis	field mint
Mentha pulegium	pennyroyal
Mentha requenii	creeping mint
Mentha spicata	spearmint
Monarda punctata	horse mint
Nepeta cataria	catnip
Origanum vulgare	wild marjoram
Polygonum amplexicaule	mountain fleece
Pycnantheum virginianum	mountain mint
Satureja montana	winter savory
Scabiosa caucasica	scabious

Solidago	goldenrod
Stachys lanata	woolly betony
Valerian officinalis	garden heliotrope
Vitis vulpina	frost grape
LATE	August)
Aster amellus	Italian aster
Aster dumosus	bushy aster
Aster laterifolius	calico aster
Aster novae-angliae	New England aster
Clematis ligusticifolia	western virgin's bower
Clematis vitalba	traveller's joy
Helenium autumnale	common sneezeweed
Liatris spicata	blazing star
Mentha × piperita	peppermint
Parthenocissus quinquefolia	Virginia creeper
Polygonum aubertii	silver fleece vine
Polygonum cuspidatum	fleece flower
Scrophularia marilandica	Maryland flower
Verbesina alternifolia	golden honey plant
VERY LATE (September)	
Aster cordifolius	blue wood aster
Aster laevis	smooth aster
Aster novi-belgii	New York aster
Aster puniceus	swamp aster
Colchicum autumnale	meadow saffron
Hedera helix	English ivy
Helianthus salicifolius	willow leaved sunflower
Helleborus niger	Christmas rose
Sedum spectabile	showy sedum

13.4.3 Calendar of trees and shrubs

Approximate season of first bloom of most trees and shrubs listed in the index:

VERY EARLY (March)	
Acer saccharinum	soft maple
Arctostaphylos tomentosa	woolly manzanita
Berberis darwinii	Darwin barberry
Berberis × stenophylla	rosemary barberry
Corylus americana	American hazel
Corylus avellana	European hazel
Daphne mezereum	February daphne
Erica carnea	spring heath
Hamemelis mollis	Chinese witch hazel
Myrica gale	sweet gale
Populus	poplar
Salix nigra	black willow
Viburnum tinus	laurestinus
EARLY (April)	
Acer circinatum	vine maple
Acer negundo	Manitoba maple
Acer platanoides	Norway maple
Acer rubrum	red maple

Alnus	alder
Amelianchier alnifolia	Saskatoon berry
Arctostaphylos columbiana	hairy manzanita
Cercis canadensis	red bud
Cornus mas	Cornelian cherry
Diervilla lonicera	honeysuckle diervilla
Lonicera xylosteum	European fly honeysuckle
Magnolia stellata	star magnolia
Pieris floribunda	pieris
Prunus armeniaca	apricot
Prunus domestica	garden plum
Prunus dulcis	flowering almond
Prunus sargentii	Sargent cherry
Prunus × *yedoensis*	Yoshino cherry
Quercus	oak
Ribes aureum	buffalo currant
Ribes rubrum	red currant
Ribes uva-crispa	English gooseberry
Rosmarinus officinalis	rosemary
Salix caprea	goat willow
Salix discolor	pussy willow
Shepherdia argentea	silver buffalo berry
Ulex americana	gorse
Ulmus americana	American elm

MODERATELY EARLY (May)

Acer campestre	hedge maple
Acer ginnala	Amur maple
Acer macrophyllum	Pacific maple
Acer palmatum	Japanese maple
Acer pseudoplatanus	sycamore
Acer tataricum	Tartary maple
Aesculus × *carnea*	red horse chestnut
Aesculus glabra	Ohio buckeye
Aesculus hippocastanum	common horse chestnut
Arbutus menziesii	Pacific madrone
Arctostaphylos uva-ursi	bearberry
Caragana arborescens	Siberian peashrub
Chaenomeles speciosa	flowering quince
Cornus stolonifera	red osier dogwood
Crataegeus laevigata	English hawthorn
Crataegeus monogyna	single seed hawthorn
Daphne cneorum	rose daphne
Deutzia × *kalmiiflora*	deutzia
Euonymus alatus	winged spindle tree
Ilex glabra	inkberry
Ilex opaca	American holly
Ledum groenlandicum	Labrador tea
Lonicera morrowii	Morrow honeysuckle
Lonicera tatarica	Tartarian honeysuckle
Magnolia acuminata	cucumber tree
Magnolia × *soulangiana*	saucer magnolia
Malus baccata	Siberian crab apple
Malus coronaria	crab apple
Malus domestica	apple
Malus pumila	common apple

Paulownia tomentosa	princess tree
Phellodendron amurense	Amur cork
Prunus avium	sweet cherry
Prunus cerasus	sour cherry
Prunus laurocerasus	cherry laurel
Prunus padus	European bird cherry
Prunus persica	peach
Prunus serrulata	Oriental cherry
Prunus spinosa	blackthorn
Prunus virginiana	western choke cherry
Ptelea trifolicata	hop tree
Pyrus communis	pear
Rhamnus cathartica	common cascara
Rhamnus purshiana	cascara buckthorn
Ribes odoratum	buffalo currant
Ribes sanguineum	winter currant
Rosa hugonis	Father Hugo rose
Rubus deliciosus	boulder raspberry
Sorbus aria	white beam
Vaccinium angustifolium	low bush blueberry
Vaccinium corymbosum	high bush cranberry
Viburnum plicatum	Japanese snowball
Viburnum prunifolium	black haw
Weigela florida	weigela
Wisteria floribunda	Japanese wisteria
Wisteria sinensis	Chinese wisteria

MID (June)

Ailanthus altissima	tree of heaven
Amorpha fruticosa	false indigo
Buddleia globosa	globe butterfly bush
Catalpa bignoides	common catalpa
Ceanothus americana	red root
Cladastrus lutea	American yellow wood
Colutea aborescens	bladder senna
Cornus alba	Tatarian dogwood
Cotoneaster horizontalis	rockspray cotoneaster
Cotoneaster microphylla	small leaved cotoneaster
Crataegeus crus-gallii	cockspur hawthorn
Cytisus albus	Portugese broom
Cytisus scorparius	Scotch broom
Deutzia scabra	fuzzy deutzia
Diospyros virginiana	common persimmon
Elaeagnus angustifolia	Russian olive
Erica cinerea	Scotch heath
Fuchsia magellanica	Magellan fuchsia
Gaultheria shallon	salal
Gleditsia triacanthos	common honey locust
Ilex aquifolium	English holly
Ilex verticillata	winterberry
Kalmia angustifolia	sheep laurel
Kalmia latifolia	mountain laurel
Laurus nobilis	laurel
Ligustrum japonicum	Japanese privet
Liriodendron tulipifera	tulip tree
Magnolia grandiflora	southern magnolia

Rhus glabra	red sumac
Rhus typhina	staghorn sumac
Robinia hispida	rose acacia
Robinia pseudoacacia	false acacia
Rosa blanda	meadow rose
Rosa multiflora	Japanese rose
Rosa rugosa	rugosa rose
Rubus spp.	blackberry
Rubus idaeus	raspberry
Rubus occidentalis	wild raspberry
Senecio greyii	senecio
Sophora viciifolia	vetch sophora
Sorbus intermedia	Swedish white beam
Symphoricarpos albus	snow berry
Tilia americana	basswood
Tilia × europaea	common lime
Tilia platyphyllos	big leaf linden
Vaccinium arboreum	tree huckleberry
Viburnum opulus	Guelder rose

MODERATELY LATE (July)

Callicarpa bodinieri	Bodinier beauty bush
Cephalanthus occidentalis	button bush
Clethra alnifolia	summersweet
Erica vagans	Cornish heath
Hydrangea	hydrangea
Koelreuteria paniculata	golden rain tree
Lavandula angustifolia	lavender
Lespedeza bicolor	shrub bush clover
Ligustrum ovalifolium	California privet
Oxydendron arboreum	sourwood
Rosa setigera	prairie rose
Rosa wichuriana	memorial rose
Rubus odoratus	flowering raspberry
Symphoricarpos orbiculatus	buckbrush
Tilia cordata	small-leaved linden
Tilia euchlora	Crimean lime
Tilia mongolica	Mongolian lime
Yucca filamentosa	Adam's needle
Yucca glauca	soapweed

LATE (August)

Aralia elata	Japanese angelica tree
Aralia spinosa	angelica tree
Elsholtzia stauntonii	heather mint
Perovskia atriplicifolia	silver sage
Sophora japonica	Japanese pagoda tree

VERY LATE (September)

Hamamelis virginiana	witch hazel
Lespedeza thunbergii	Thunberg lespedeza
Vitex negundo cv. *heterophylla*	cut-leaved chaste tree

TABLE 19. Some domestic crops visited by honeybees, indicating approximate month and duration of bloom; value as nectar or pollen source; which crops are pollinated by honeybees; and the quality of the honey produced (where known). N, nectar; P, pollen; HP, honey potential[a]; D, honeydew; x, affirmative; *, see index for more information; ( ), indicates minor importance for N, P or D.

Common name	Scientific name	Month of bloom &/or duration	Value	Honeybees known to increase seed &/or fruit set	Honey	Notes
almond *	Prunus amygdallus	late Apr	NP	x	bitter quality; poor	HP1/2 (for Prunus spp.)
apple *	Malus spp.	late Apr	NP	x	light in colour;	HP 1/2 (for Malus spp.)
apricot *	Prunus armeniaca	late Apr	NP	x	similar to apple	HP2
artichoke	Cynara scolymus		NP	x?		
asparagus	Asparagus officinalis	Jun-Jul	NP	x	light to medium amber in colour; usually inferior	H yield of 0.2 to 1 kg/ colony/day
bean, broad *	Vicia faba	Jun-Aug	NP	?	light to dark amber, granulation rapid; may yield honeydew	HP3
bean, common	Phaseolus vulgaris		(NP)			visited only sparingly
bean, lima	Phaseolus lunatus	H flow 1 week	N(P?)		water white; mild; characteristic yeasty flavour	intense N flow; H yield 27 kg/ colony/season
beet	Beta vulgaris		(N?)P			P collected if little else available

TABLE 19. Continued

Common name	Scientific name	Month of bloom &/or duration	Value	Honeybees known to increase seed &/or fruit set	Honey	Notes
blackberry *	Rubus fruticosa & spp.	Jun-frost (peak Aug)	NP	x	classed as white to light amber; granulation slow	HP1/2
blueberry, high bush *	Vaccinium corymbosum	late May-early June	NP	x	light in colour; may granulate hard; excellent flavour	H yields up to 40 kg/colony/season
carrot	Daucus carota	1 month	NP	x	dark amber; flavour strong	high yields on heavy soils and temps. above 27°C
celery	Apium graveolana	ea. flower lasts a few days	N	x	fresh H tastes of celery	abundant yields
cherry *	Prunus	late Apr-mid-May	NP	x	light; granulates quickly; fine grain	HP1/2 (for Prunus spp. only)
chives	Allium schoenoprasum	Jun	NP	x	light amber; good quality; similar to onion	HP3 (for Allium spp.)
cole crops	Brassica	ea. flower lasts a few days	NP	x?	usually good quality	a rich source of H
corn *	Zea mays	Jul-Aug	DP		honeydew H sometimes produced; quality varies; dark; coarse	D gathered in leaf axils or where tissue damaged
crab apple *	Malus	late May	NP	x	similar to apple	HP1/2 (for Malus spp.)

TABLE 19. Continued

Common name	Scientific name	Month of bloom &/or duration	Value	Honeybees known to increase seed &/or fruit set	Honey	Notes
cranberry *	Vaccinium macrocarpon	Apr	(N)P	x	light amber in colour; flavour fine	H seldom obtained
cucumber	Cucumis sativus		NP	x	pale yellow in colour; strong flavour initially	HP2 (for Cucumis spp.)
currant, red & black *	Ribes rubrum Ribes nigrum	Apr-May	NP	x?	unknown in pure form	HP4 (Ribes rubrum); R. nigrum alternate host for white pine blister rust
gooseberry *	Ribes uva-crispa	Apr	N(P)	x	medium coloured; flavour excellent	HP3
grape *	Vitis vinifera		DNP	x	honeydew may be collected; H from floral N may be reddish	honeybees unable to puncture grape skin to suck juice
leek	Allium porrum	Jun	N(P)	x	similar to onion.	HP3 (for Allium spp.)
melon	Cucumis spp.		NP	x	light amber; good	HP2 (for Cucumis spp.)
onion *	Allium cepa	Jun: 2-3 weeks	N(P)	x	light amber; good quality; oniony flavour disappears	HP3

TABLE 19. Continued

Common name	Scientific name	Month of bloom &/or duration	Value	Honeybees known to increase seed &/or fruit set	Honey	Notes
parsnip	Pastinaca sativa	Jun-Jul	N(P?)	?	inferior to celery H; poor quality	yields abundant N
pea	Psium sativum		N(P?)	x?	crystal white in colour, very strong taste of peas	not highly attractive to honeybees
peach *	Prunus persica	early May	NP	x	similar to apple	HP1/2: P is an excellent protein source
pear *	Pyrus spp.	early May	NP	x	similar to apple	few, if any, reports of pear H
plum *	Prunus domestica	late Apr	NP	x	similar to apple	HP2
squash	Cucurbita spp.		NP	x	good quality; surplus obtained	N freely secreted; P highly attractive to honeybees
strawberry	Fragus spp.	Jun	N(P)	x	unknown in pure form	not consistently attractive to honeybees
turnip *	Brassica spp.		N(P)	x?	good quality	N freely secreted; attractive source

a See 4.7 for explanation of honey potential.

14. POSTSCRIPT

This book is my first attempt to write anything longer than a term paper, and, having finished, I am sure that I have been overly ambitious. The topic is far larger and more fascinating than ever I had imagined when I began about 4 or 5 years ago. Unless I had made a disciplined effort to put down my pen, I would have carried on indefinitely. I cannot say that it is not a relief to finish!

There are some aspects of the book which need improving. These are probably as obvious to the reader as they are to me. First of all, an enormous drawback is the lack of illustration. It is hoped that it might be possible to produce a series of numbered slide strips for hand viewing or a set of colour plates, of many of the species listed in the text. These illustrations would be offered for sale as an accompaniment to the book. Unfortunately, simultaneous production of the book and a set of slides or colour plates was not financially possible.

Also, several sections need expanding or refining. In particular, the section on Canadian beekeeping needs to be expanded to cover more geographic regions; Table 14 could probably be refined further; more French common names could be given; further practical observations by beekeepers and researchers might be compiled and included; and the section on honeydew could be made more relevant to Canada.

Finally, I must apologize for omissions of plant species that should have been included and for any inadvertent errors or inconsistencies in the text. It would be impossible to include every species visited by honeybees in Canada, but I hope that most are noted in this book.

Owing to the scarcity of published Canadian reference material, contemporary[a] or historical, a future improved version would largely depend on the extent and quality of feedback from beekeepers, apiculturalists, botanists and others.

> "... Then, when the late year wastes,
> When night falls early and the noon is dulled
>   And the last warm days are over,
> Unlock the store and to your table bring
> Essence of every blossom of the spring.
> And if, when the wind has never ceased to blow
> All night, you wake to roofs and trees becalmed
>   In the level wastes of snow,
> Bring out the Lime-tree-honey, the embalmed
> Soul of a lost July, or Heather-spiced
> Brown-gleaming comb wherein sleeps crystallized
> All the hot perfume of the heathery slope.
> And tasting, remembering, live in hope."

Martin Armstrong, *Honey Harvest*

---

[a]   A welcome exception is *Ontario Nectar Trees, Shrubs and Herbs* by Larson and Shuel published in 1987.

15. REFERENCES

15.1 References cited

1. Adams, R.J.; Smith, M.V.; Townsend, G.F. (1979) Identification of honey sources by pollen analysis of nectar from the hive. *J. apic. Res.* 18 (4): 292-297
2. Adey, M.; Walker, P.; Walker. P.T. (1986) Pest control safe for bees: a manual and directory for the tropics and subtropics. *London, UK.: International Bee Research Association*
3. Alex, J.F.; Switzer, C.M. (undated) Ontario Weeds. *Agdex 640, Publ. 505, Ont. Min. Agr. and Food, Toronto*
4. Arnason, A.P. (1964) Recent studies in Canada of crop pollination by insects. *Bee Wld 47 (1) (Suppl.) (II Int'l Symp. on Pollination)*, 107-124
5. Arnold Arboretum (1971) Hardiness zones of the United States and Canada. *Arnold Arboretum, Harvard University, Mass. End-papers in* Wyman's Gardening Encylopaedia, D. Wyman. *New York: MacMillan.*
6. Atkins, E.L. (1975) Injury to bees by poisoning. *Pp.* 690-691 *in* The Hive and the Honeybee, *ed.* Dadant and Sons. *Hamilton, Ill.: Dadant*
7. Barker, R.J. (1978) Poisoning by plants. *In* Honeybee Pests, Predators and Diseases, *ed.* R.J. Morse. *Ithaca, N.Y.: Cornell University Press*
8. Barker, R.J.; Lehner, Y. (1976) Galactose, a sugar toxic to honeybees found in the exudate of tulip flowers. *Apidologie* 7 (2): 109-111
9. Bendek, P.; Komlodi, J.; Penner, J.; Wilheim, E. (1972) Insect pollination of oilseed rape (*Brassica napus* L.). *Novenytermeles* 21 (3): 255-269
10. Burnside, C.E.; Vansell, G.H. (1936) Plant poisoning of bees. *U.S.D.A., Bureau of Entomology and Plant Quarantine, Washington*
11. Corner, J.; Lapkins, K.O.; Arrand, J.C. (1964) Orchard and honeybee management in planned fruit pollination. *Apiary Circ. 14, B.C. Dept. Agric., Victoria, British Columbia.* 18 pp.
12. Couture, J.M. (1959) Beekeeping in Quebec. *Gleanings in Bee Culture* 87 (8): 462-467
13. Crane, E. (1966) Canadian bee journey. *Bee Wld* 47 (2/4): 55-65, 132-148
14. Crane, E. [ed.] (1976) Honey: a Comprehensive Survey. *London, UK: Heinemann.* 608 pp.
15. Crane E. (1976) The world's honey production. *Pp.* 115-153 *in* Honey: a Comprehensive Survey, *ed.* E. Crane. *London, UK: Heinemann*
16. Crane, E. (1976) The flowers honey comes from. *Pp.* 3-76 *in* Honey: a Comprehensive Survey, *ed.* E. Crane. *London, UK: Heinemann*
17. Crane, E. Walker, P.; Day, R. (1984) Directory of Important World Honey Sources. *London, UK: International Bee Research Association.* 384 pp.
18. Crane, E.; Walker, P. (1985) Important honeydew sources and their honeys. *Bee Wld* 66 (3): 105-112
19. Dadant and Sons [eds.] (1975) The Hive and the Honeybee. *Hamilton, Ill.: Dadant.* 740 pp.
20. Davidson, J. (Undated) List of British Columbia flora. *Vancouver, Canada: University of British Columbia.* 4 pp.
21. Ewert, R. (1929) Die befruchtung der cruciferenblute durch die bienen. *Archiv. Bienenk.* 10: 310-312
22. Feller-Demalsy, M.J.; Lamontagne, Y. (1979) Analyse pollinique des miel du Quebec. *Apidologie* 10 (4): 313-339
23. Fernald, M.L. (1950) Gray's Manual of Botany. *New York: American Book Company.* 1632 pp.

24. Free, J.B. (1976) Insect Pollination of Crops. *London, UK: Academic Press.* 544 pp.

25. Free, J.B.; Ferguson, A.W. (1983) Foraging behaviour of honeybees on oilseed rape. *Bee Wld* 64 (1): 22-24

26. Free, J.B.; Nuttall, P.M. (1968) The pollination of oilseed rape (*Brassica napus* L.) and the behaviour of bees on the crop. *J. agric. Sci., Camb.* 71: 91-94

27. von Frisch, K. (1950) Bees: their Vision, Chemical Senses, and Language. *London, UK: Johnathan Cape.* 125 pp.

28. Goltz, L. (1986) Honey and pollen plants, part 1: milkweeds. *Am. Bee J.* 126 (9): 601-602

29. Gorbach, G. (1952) Biene und Blatthonig. *Imker,* Berlin 4 (5): 138-142

30. Hayes, B. (1978) The sourwood tree and its honey - Distinctive and unique. *Am. Bee J.* 118 (2): 86-87

31. Hayes, B. (1979) Purple loosestrife, the wetlands honey plant. *Am. Bee J.* 119 (5): 382-383

32. Holmes, F.O. (1980) Oaks and oak relatives as nectar sources. *Glean. in Bee Cult..*

33. Howes, F.N. (1949) Poisoning from honey. *Food Manu.* 24: 459-463

34. Howes, F.N. (1979) Sources from poisonous honey. *Kew Bull., London* 2: 167-171

35. Howes, F.N. (1979) Plants and Beekeeping. *London, UK: Faber and Faber.* 236 pp.

36. Int'l Bee Res. Assoc. (1981) Garden Plants Valuable to Bees. *London, UK: International Bee Research Association.* 52 pp.

37. Int'l Union of Plant Sci. (1980) International code of nomenclature for cultivated plants. *Regnum Vegetabile, The Hague* 104: 1-32

38. Int'l Union of Plant Sci. (1983) International code of botanical nomenclature. *Regnum Vegetabile, The Hague* 111: 1-472

39. Jamieson, C.A. (1958) Facts about beekeeping in Canada. *Bee Wld* 39 (9): 232-236

40. Johnson, K.W. et al. (1976) Nectar for your bees. *Your Forests, Ont. Min. Nat. Res., Forest Management Branch* 9: 24-26

41. Kamler, F. (1983) The response of selected winter rape cultivars to the pollination of honeybees. *Vyroba* 29 (3): 225-234

42. Karmo, E.A. (1958) Honeybees as an aid in orchard and blueberry pollination in Nova Scotia. *Proc. Xth Int. Congr. Ent. 1956:* 955-959

43. Kervlit, J. (1981) Analysis of a toxic rhododendron honey. *J. apic. Res.* 20 (4): 249-253

44. Kohanawa, M. (1956) Pharmacological studies in rhodotoxin isolated from *Rhododendron hyemanthes. Jap. J. Pharmacology* 6: 46-56

45. Langridge D.F.; Goodman, R.D. (1975) A study on pollination (*Brassica campestris* L.). *Aust. J. exp. Agric. Anim. Husb.* 15: 285-288

46. Langridge, D.F.; Goodman, R.D. (1982) Honeybee pollination of oilseed rape, cultivar 'Midas'. *Aust. J. exp. Agric. Anim. Husb.* 22: 124-126

47. Latif, A.; Qayyum, A.; Abbas, M. (1960) The role of *Apis indica* in the pollination of 'Toria' and 'Sarson' (*Brassica campestris* var. *toria* and *dischotoma). Bee Wld* 33 (11): 283-286

48. Leach, D.G. (1965) Ancient curse of rhododendron honey. *Garden J.* 16: 215-217, 239

49. Liberty Hyde Bailey Hortorium (1976) Hortus Third, Concise Dictionary of Plants Cultivated in the United States and Canada. *Ithaca, N.Y.: State College of Agric. Cornell University*

50. Lovell, H. (1966) Honey Plants Manual. *Medina, Ohio: A.I. Root.* 64 pp.

51. Mansfeld, R. (1986) Verseichnis landwirtschafhcher und gartnerischer kulturpflanzen (ohne zierpflanzen). B and I bis 4. *Berlin: Springer Verlag.* 1998 pp.

52. Maurizo, A. (1970) How bees make honey. *Pp.* 77-97 *in* Honey: a Comprehensive Survey, ed. E. Crane. *London, UK: Heinemann*

53. McCutcheon, D.M. (1958) Beekeeping in Saskatchewan. *Glean. in Bee Cult.* 86 (6): 329-331

54. McGregor, S.E. (1976) Insect Pollination of Cultivated Crop Plants. *Agric. Handbook No. 496, US Dep. Agric. Washington, D.C.* 411 pp.

55. Melville, P. (1949) The limes as amenity trees. *Kew Bull. No. 2, London, UK*

56. Milum, V.G. (1957) Illinois Honey and Pollen Plants. *Dept. of Hort., University of Ill., Urbana, Ill.*

57. Mitchener, A.T. (1948) Nectar and pollen producing plants of Manitoba. *Sci. Agric., Ottawa.* 28: 473-480

58. Mohammad, A. (1935) Pollination studies in toria *Brassica napus* L. var. *dichotomas,* Prain) and sarson (*Brassica campestris* L. var. *sarson* Prain). *Cited in* Honey bees and canola, a perfect pair, Mohr, N.A. *Can. Beekeep.* 13 (1): 7-8

59. Mohr, N.A. (1986) Honey bees and canola, a perfect pair. *Can. Beekeep.* 13 (1): 7-8

60. Mohr, N.A.; Jay, S.C. (1986). The foraging behaviour of honeybees on canola (*Brassica campestris* L. and *Brassica napus* L.). *From* Honeybees and canola, a perfect pair. Mohr, N.A. *Can. Beekeep.* 13 (1): 7-8. *In Press*

61. Mohr, N.A.; Jay S.C. (1986) Nectar production of selected cultivars of *Brassica campestris* L. and *Brassica napus* L. *From* Honeybees and canola, a perfect pair. Mohr, N.A. *Can. Beekeep.* 13 (1): 7-8. *In Press*

62. Morrison, W.C. (1957) Woody honey plants for roadside plantings in New Jersey. *Circ. No. 403, State of N.J., Dept. of Agric.* 22 pp.

63. Morse, R.J. [ed.] (1978) Honeybee Pests, Predators and Diseases. *Ithaca, N.Y.: Cornell University Press.* 430 pp.

64. Morton, J.F. (1958) Ornamental plants with poisonous properties. *Proc. of the Fla. State Hort. Soc.* 71: 272-277

65. Munz, P.A.; Keck, D.D. (1959) A California Flora. *Berkeley and Los Angeles: University of California Press.* 1681 pp.

66. Murrell, D.C.; Szabo, T. (1981) Pollen collection by honeybees at Beaverlodge, Alberta. *Am. Bee J.* 121 (12): 885-888

67. Nye, W.P. (1971) Nectar and pollen plants of Utah. *Utah State University Monograph Series XVIII (3).* 79 pp.

68. Oldszowy, D.R. (1977) Of bees, rhododendron and honey. *Am. Bee J.* 117 (8): 498-500

69. Olason, G. (1960) Self incompatibility and outcrossing in rape and white mustard. *Hereditas* 46: 241-252

70. Ouller, C.A.; Sherk, L.C. (1973) Map of plant hardiness zones in Canada. *Agric. Can. Publ. 5003, Ottawa.*

71. Palmer-Jones, T. (1947) A recent outbreak of honey poisoning. *N.Z. Jl. Sci. Technol.* 29 (3) (Sec. 3): 107-134

72. Pellet, F.C. (1977) American Honey Plants. *Hamilton, Ill.: Dadant and Sons.* 467 pp.

73. Prov. of British Columbia (1938) Bee culture in British Columbia. *Bull. 92, Dept. of Agric., Victoria.* 56 pp.

74. Purseglove, J.W. (1968) Tropical Crops, Dicotyledons I. *London, UK: Longmans* 332 pp.

75. Radchenko, T.G. (1964) The influence of pollination of the crop and the quality of winter rape. *Bolzhil'nitstro* 12: 68-74

76. Rachman, K.A. (1940) Insect pollinators of toria (*Brassica napus* L. var. *dichotoma* Prain) and sarson (*Brassica campestris* L. var. *sarson*). *Indian J. agric. Sci.* 10: 422-447

77. Rao, G.M.; Suryanarayana, M.C.; Thakar, V. (1980) Bees can boost oilseed production. *Indian Fmg.* 29 (11): 15-26

78. Robinson, F.A.; Oertel, E. (1975) Sources of nectar and pollen. *Pp.* 283-302 *in* The Hive and the Honeybee, *ed.* Dadant and Sons. *Hamilton, Ill.: Dadant and Sons*

79. Scullen H.P.; Vansell, G.A. (1947) Nectar and pollen plants of Oregon. *Bull 412, Agric. Expl. Sta., Oregon State College, Corvallis, Oregon*

80. Selwyn, H.H. (1949) Save the basswoods. *Can. Bee J.* 57 (11): 4-5

81. Sherk, L.C.; Buckley, R. (1974) Ornamental shrubs for Canada. *Minister of Supply and Services, Ottawa.* 187 pp.

82. Shuel, R.W. (1975) The production of nectar. *Pp.* 265-278 *in* The Hive and the Honeybee, *ed.* Dadant and Sons. *Hamilton, Ill.: Dadant and Sons*

83. Stanley, R.G.; Linskens, H.F. (1974) Pollen: biology, biochemistry and management. *Berlin: Springer-Verlag.* 307 pp.

84. Stroempl, G. (1977) Basswoods and lindens for honey production. *Am. Bee J.*

85. Szabo, T.I. (1980) Nectar secretion by 28 varieties and breeder lines of two species of rapeseed (*Brassica napus* and *Brassica campestris*) *Am. Bee J.* 122 (9): 645-647

86. Szabo, T.I (1985) Variability of flower, nectar, pollen and seed production in some Canadian canola (rapeseed) varieties. *Am. Bee J.* 125 (5): 351-353

87. Terry, H.A. (1872) Beekeeper's Journal, March. *Pp.* 185-186 *in* American Honey Plants, Pellet, F.C. *Hamilton, Ill.: Dadant and Sons*

88. Townsend, G.F. (1966) The crop without a surplus. *Can. Bee J.* 64 (2): 19-21

89. US Dep. Agric., F.A.S. (1985) Sugar, Molasses and Honey: World Honey Situation. *Horticultural and Tropical Division, US Dep. of Agric., Washington. D.C.*

90. Vansell, G.H. (1941) Nectar and pollen plants of California. *Bull. 517, University of California, Berkley.* 76 pp.

91. Walsh, R.S. (1978) Nectar and Pollen Sources of New Zealand. *Beekeep. Assoc. of N.Z., Wellington, N.Z.* 59 pp.

92. White, J.W. (1975) Honey. *Pp.* 491-530 *in* The Hive and the Honeybee, *ed.* Dadant and Sons. *Hamilton, Ill.: Dadant and Sons*

93. Williams, H. (1980) Oilseed rape and beekeeping, particularly in Britain. *Bee Wld* 61 (4): 141-153

94. Wilson, W.T.; Moffatt, J.O.; Hamilton, J.D. (1958) Nectar and pollen plants of Colorado. *Bull. 503-505, Colo. Expl. Sta., Fort Collins, Ore.* 71 pp.

95. Wyman, D. (1971) Wyman's Gardening Encyclopaedia. *New York: MacMillan.* 1221 pp.

96. Zander, E. (1951) Ras and bienen. *Z. Bienenforsch* 1 (8): 135-140.

15.2 Further references

The following references include both technical and popular literature in horticulture or apiculture that are not cited in the text. This is by no means a comprehensive list of any kind, but serves only to add the names of some useful references not previously mentioned. Only a few of the many excellent and practical Provincial and Federal publications could be given here. A complete list of government publications may be obtained from District Departments of Agriculture, Research Stations, or from Ottawa.

1. Agric. Can. (1971) What you should know about oilseed crops. *Publ. 1448, Agric. Can., Ottawa.* 11 pp.

2. B.C. Dep. of Agric. (1975) Pollination and Fruit Set in Tree Fruits. *B.C. Dep. of Agric., Victoria.* 20 pp.

3. Buckley, A.R. (1986) Trees and Shrubs of the Dominion Arboretum, *Publ. 1697, Res. Br., Agric. Can.* 237 pp.

4. Chipman, E.W. (1975) Growing Savory Herbs. *Publ. 1158, Agric. Can., Ottawa.* 15 pp.

5. Cole, T.J. (1979) A Checklist of Ornamental Trees for Canada. *Publ. 1343, Agric. Can., Ottawa.* 28 pp.

6. Cole, T.J. (1980) Ground Covers and Climbing Plants. *Publ. 1698, Agric. Can., Ottawa.* 38 pp.

7. Crane, E.; Walker, P. (1983) The Impact of Pest Management on Bees and Pollination. *Trop. Dev. and Res. Inst., London.* 232 pp.

8. Dedio, W.; Hoes, J.A. (Undated) Sunflower Seed Crops. *Publ. 1687, Agric. Can., Ottawa.* 31 pp.

9. Faergi, K.; Iverson, B.; Waterbolk, J. and H.T. (1966) Textbook of Pollen Analysis. *Munkgard, Copenhagen.* 237 pp.

10. Fleming, R.A.; Hammersma, B. (Undated) Herbaceous Perennials. *Pub. 358, Ont. Min. of Agric. and Food.* 20 pp.

11. Goplen, B.P.; Gross, A.T. (1978) Sweet Clover Production in Western Canada. *Publ. 1613, Agric. Can., Ottawa.* 15 pp.

12. Hall, I.V.; Jackson, L.P. *et al.* (1979) Lowbush blueberry production. *Publ. 1477, Agric.*

Can., Ottawa. 49 pp.

13. Hanna, M.R.; Cooke, D.A. *et al.* (1977) Sainfoin for Western Canada. *Publ. 1470, Agric. Can., Ottawa.* 18 pp.

14. Hodges, D. (Undated) Flowers for Bees, Month by Month. *Int. Bee Res. Assoc., London.* 10 pp.

15. Littler, A.E. (1974) Holly Culture in British Columbia. *B.C. Dep of Agric., Victoria.* 11 pp.

16. Oertel, E. (1939) Honey and Pollen Plants of the United States. *Circ. 554, US Dep. Agric., Washington, D.C.* 64 pp.

17. Oliver, R.W. (1967) Hedges for Canadian Gardens. *Publ. 899, Agric. Can., Ottawa.* 22 pp.

18. Oliver, R.W. (1975) Descriptive Notes on Herbaceous Perennials for Canadian Gardens. *Publ. 968, Agric. Can., Ottawa.* 37 pp.

19. Oliver, R.W. (1978) Annual Flowers. *Publ. 796, Agric. Can., Ottawa.* 32 pp.

20. Oliver, R.W. (1978) Trees for Ornamental Planting. *Publ. 995, Agric. Can., Ottawa.* 31 pp.

21. Proctor, M.; Yeo, P. (1979) The Pollination of Flowers. *Collins, London.* 418 pp.

22. Root, A.I.; Root, E.R., (1935) The ABC and XYZ of Bee Culture. *A.I. Root, Medina, Ohio.* 814 pp.

23. Traynor, J (1966). Increasing pollinating efficiency of honeybees. *Bee Wld* 47 (3): 101-110.

24. Verey, R. (1981) The Scented Garden. *Van Nostrand Reinhold, N.Y.* 168 pp.

16. INDEX OF NAMES

16.1 Common Names

foxglove 22
fragrant giant hyssop 36
framboisier 103
framboisier d'Amerique 103
framboisier noir 103
fraxinella 46
fraxinelle 46
French honeysuckle 49,107
French tamarisk 107
fresillon 87
frigoule 64
frost flower 39
frost grape 67
fruit trees 4
fuchsia
    hardy 82
    Magellan 82
Fuji cherry 93
fuller's teasel 22,124
furze 111
fusain ailé 82
fuzzy deutzia 80
gadelier 100
gaillardia 7
    painted 24
gainier 76
gale
    sweet 91
gale odorant 91
gallberry 84
garden balsam 25
garden heliotrope 66
garden hollyhock 16
garden hyssop 50
garden plum 93,94
garden sage 60
garden thyme 64
garland flower 80
garlic 36,37
gasplant 46
gattelier 113
gayfeather
    Kansas 52
    spike 52
gaylussacia 4
genêt à balais 80
Geneva bugle 36
geranium
    blood red 48
    meadow 48
géranium de près 48
géranium livide 48
gesse tubereuse 51
giant hyssop 4,36
giant onion 37

gilia
    globe 24
giroflée jaune 20
giroflée violier 20
gladiolus 48
glads 48
glaieul 48
globe butterfly bush 75
globe candy tuft 25
globe gilia 24
globe thistle 46
gloire de neige 43
glory of the snow 43
glycine de Chine 114
glycine du Japon 113
goat willow 104,120
gogoune 93
golden cleome 21
golden currant 99,100
golden flax 52
golden gram 34
golden honey plant 2,67
golden rain tree 85
goldenrod 4,39,62,138
    Canada 62
    Canadian 7,62
    western 7,62,63
goldentuft 42
gooseberry
    English 100
gorse 111
gourdin d'Hercule 73
grand érable 70
grand sauge 60
grand soleil 25
grand trèfle 54
grandes brimbelles des marais 111
grape
    bear's 74
    chicken 67
    frost 67
    mountain 89
    Oregon 89
    winter 67
gray willow 104
great globe thistle 46
Greek anemone 38
Greek valerian 59
grelot blanc 51
groseillier 99
groseillier des buffles 100
groseillier épineux 100
groseillier rouge 100
groseillier sanguin 100
ground hemlock 108

red haw 79
red horse chestnut 71
red maple 69,70,76,77
red osier dogwood 78
red sumac 99
red top grasses 65
red whortleberry 113
redbud 76
 eastern 76
redroot 76
redvein enkianthus 81
renoncule 59
reseda odorante 31
rhododendron 99
 Pontic 99,125
 tree 125
river locust 73
river maple 71
robinier rose 100
rock cotoneaster 78
rock cress
 purple 41
 wall 38
rock madwort 42
rockspray cotoneaster 79
Rocky Mountain bee plant 2,7,21
Rocky Mountain flowering raspberry 103
Rocky Mountain raspberry 103
romarin 102
ronce 103
rose 7,90,101
 Christmas 50
 Corsican 50
 Guelder 113
 hybrid tea 101
 Japanese 102
 Lenten 50
 meadow 101,102
 memorial 102
 prairie 102
 robinier 100
 rugosa 102
 Whitten 113
rose acacia 100
rose clarkia 20
rose d'hiver 50
rose daphne 80
rose de Gueldre 113
rose de Noël 50
rose tremière 16
rosebay willowherb 46
rosemary 102
 bog 73
 marsh 52
 wild 86

rosemary barberry 74
rosmarin 102
rowan 106
rowan tree 106
rudbeckie 31
rudbeckie hérissée 31
rudbeckie lacinée 60
rue
 meadow 64
rugosa rose 102
runner bean 31
Russian globe thistle 46
Russian knapweed 19
Russian olive 81
Russian thistle 5
safflower 5,19
saffron 19
 bastard 19
 false 19
 meadow 45
saffron bâtarde 19
sage 7,60
 black 60
 clary 31
 common 60
 garden 60
 meadow 60
 scarlet 32
 silver 91
 wood 60
sainfoin 5,17,36,54,57
sainfoin d'Espagne 49
Sakhalin knotweed 59
salal 83
salicaire 53
sallow 104
salmonberry 104
salsola 5
salt cedar 118
salvation Jane 23
salvia 60
sand cherry 93,94
sanguinaire 48
sapin 68
sapin argenté 68
sapin blanc 68
sapin commun 68
Sargent cherry 93,95
sarrasin 23
sarrasin de Tartarie 24
sarriette 32
sarriette annuelle 61
sarriette vivace 61
Saskatoon 72
saucer magnolia 89

16.2 Scientific Names

Abies 8,68
Abies alba 68,68,121,124
Abies concolor 68,75
Abies pectinata 68
Acer 68,5,69,70,71
Acer campestre 69
Acer circinatum 69
Acer ginnala 69
Acer macrophyllum 69
Acer negundo 70,69
Acer palmatum 70,69
Acer platanoides 70,120
Acer pseudoplatanus 70,69,120
Acer rubrum 70,9,69,76,77
Acer saccharinum 71,69
Acer saccharum 69
Aconitum 124
Actinomeris alternifolia 67
Actinomeris squarrosa 67
Aesclepias 124
Aesculus 71,72°
Aesculus californica 71,124
Aesculus × carnea 71
Aesculus glabra 71
Aesculus hippocastanum 71
Aesculus parvia 71
Agastache anethiodora 36
Agastache foeniculum 36,2,50
Agastache nepetoides 2,4,36
Ailanthus altissima 72
Ajuga genevensis 36
Ajuga repens 36
Ajuga reptans 36
Alcea ficifolia 16
Alcea rosea 16,7,9,37
Allium 36,37
Allium ameloprasum 37
Allium cepa 36
Allium giganteum 37
Allium porrum 37
Allium sativum 37
Allium schoenoprasum 37
Alnus 72,4,8
Alnus incana 72
Alnus rugosa 72
Althaea ficifolia 16
Althaea officinalis 37
Althaea rosea 9,16
Alyssum arduini 42
Alyssum maritimum 27

Alyssum orientale 42
Alyssum saxatile 42
Ambrosia 5
Amelanchier alnifolia 72
Amelanchier canadensis 72
Amorpha canescens 73
Amorpha fruticosa 73
Ampelopsis quinquefolia 58
Amygdalis communis 94
Anaphalis margaritacea 37
Anchusa azurea 37
Anchusa italica 37
Anchusa officinalis 37
Andromeda 73,122,124
Andromeda glaucophylla 73
Anemone blanda 38
Anemone nemorosa 38
Anemone patens 38
Apocynum androsaemifolium 4
Arabis caucasica 38
Arabis glabra 124
Aralia elata 73
Aralia spinosa 73
Arbutus menziesii 73
Arctostaphylos 94
Arctostaphylos columbiana 74
Arctostaphylos manzanita 74
Arctostaphylos uva-ursi 74
Armeria maritima 38
Armeria vulgaris 38
Asclepias 38,5
Asclepias incarnata 39
Asclepias speciosa 39
Asclepias syriaca 39
Asclepias tuberosa 39
Asclepias verticillata 39
Aster 39,4,40,41,62
Aster adscendens 40
Aster amellus 40
Aster azureus 39
Aster canadensis 62
Aster ciliolatus 39
Aster cordifolius 40
Aster dumosus 40
Aster ericoides 39
Aster falcatus 39
Aster foliaceus 40
Aster incarnata 39
Aster italica 37
Aster johannensis 40

Aster laevis 40,7,14
Aster lateriflorus 40
Aster lepidus 62
Aster macrophyllus 39
Aster novae-angliae 41,7
Aster novi-belgii 41
Aster puniceus 41
Aster radula 39
Aster sagittifolius 39
Aster sedifolius 39
Aster sericeus 39
Aster simplex 40
Aster speciosa 39
Aster tardifolius 39
Aster tradescantii 39
Aster umbellatus 39
Aster undulatus 39
Aster vimineus 39
Astralagus cicer 41
Astralagus sinicus 41
Aubrieta deltoidea 41
Aurina saxatilis 42
Ballota nigra 53
Baptisia australis 42
Baptisia tinctoria 42
Berberis 74,89
Berberis aquifolium 89
Berberis darwinii 74
Berberis empetrifolia 74
Berberis × stenophylla 74
Betulus 8
Bidens 21
Borago officinalis 16
Brassica 16,5,7,32,138
Brassica alba 32
Brassica campestris 17,18
Brassica napella 18
Brassica napus 17,18,19,32
Brassica nigra 17
Brassica oleifera 17
Brassica praecox 17
Brassica rapa 18,17,19
Brassica rapoeuropea 18
Buddleia 75
Buddleia davidii 75
Buddleia globosa 75
Callicarpa bodinieri 75
Callicarpa giraldiana 75
Calluna vulgaris 75
Calocedrus decurrens 75,68,121
Caltha palustris 42,124
Calystegia sepium 21
Camassia 42
Camassia cusickii 42
Camassia quamash 42

Camassia scilloides 42
Campanula 19
Campanula carpatica 42,19
Campanula medium 19
Campanula persicifolia 19
Campanula pyramidalis 19
Caragana aborescens 75
Caragana pygmaea 75
Cardiaca vulgaris 51
Carthamnus tinctorius 19,5
Castanea 8
Catalpa bignonioides 76
Ceanothus 76
Ceanothus americanus 76
Ceanothus prostratus 76
Ceanothus sanguineus 76
Ceanothus velutinus 76
Cedrus 8
Celastrus scandens 43
Centaurea 19,20,43
Centaurea cyanus 19
Centaurea dealbata 43
Centaurea moschata 20
Centaurea nigra 19
Centaurea repens 4,19
Cephalanthus occidentalis 76
Cercis canadensis 76
Chaenomeles japonica 77
Chaenomeles speciosa 77
Chamaenerion angustifolium 46
Cheiranthus allioni 20
Chieranthus cheiri 20
Chionodoxa luciliae 43
Cichorium intybus 43,4,7
Cirsium 43,5
Cirsium arvense 43
Cirsium crispus 43
Cladastrus lutea 77
Clarkia elegans 20
Clarkia unguiculata 20
Claytonia virginica 44
Clematis 44
Clematis armandii 44
Clematis ligusticifolia 44
Clematis montana 44
Clematis paniculata 45
Clematis virginiana 44
Clematis vitalba 45
Cleome arborea 20
Cleome giganteum 20
Cleome grandis 20
Cleome hasslerana 20
Cleome lutea 21
Cleome serrulata 21,2,7
Cleome spinosa 20

Clethra alnifolia 77,76
Clinopodium vulgare 61
Colchicum autumnale 45
Colutea arborescens 77
Convolvulus arvensis 21
Convolvulus sepium 21
Convolvulus tricolor 21
Coreopsis 21
Coriandrum sativum 22
Cornus 78
Cornus alba 78
Cornus mas 78
Cornus sericea 78
Cornus stolonifera 78
Coronilla varia 45
Corylus 8
Corylus americana 78
Corylus avellana 78
Cosmos bipinnatus 22
Cotoneaster 78,79
Cotoneaster conspicuous 78
Cotoneaster franchetti 78
Cotoneaster frigidus 78
Cotoneaster horizontalis 79
Cotoneaster microphyllus 79
Cotoneaster simonsii 79
Crataegus 79,4
Crataegus crus-galli 79
Crataegus laevigata 79
Crataegus oxyacantha 79
Crataegus punctata 79
Crocus 45,8,123
Cucumis sativa 7
Cynoglossum officinale 4
Cyrilla racemiflora 124
Cytisus 80
Cytisus alba 80
Cytisus scoparius 80
Dahlia 45
Daphne cneorum 80
Daphne mezereum 124
Datura stromonium 124
Delphinium consoida 124
Deutzia 80
Deutzia × kalmiiflora 80
Deutzia parviflora 80
Deutzia purpurascens 80
Dictamnus albus 46
Diervilla lonicera 80
Diervilla sessifolia 81
Digitalis purpurea 22,124
Diospyros virginiana 81
Dipsacus fullonum 22
Dipsacus lacinatus 22
Dipsacus sativus 22,124

Dipsacus sylvestris 22
Doria canadensis 62
Doronicum plantagineum 46
Dracocephalum moldavica 22
Echinops exaltatus 46
Echinops ritro 46
Echinops sphaerocephalus 46,2
Echium lycopsis 23
Echium plantagineum 23
Echium vulgare 23,4,5
Elaeagnus angustifolia 81
Elsholtzia crispa 81
Elsholtzia cristata 81
Elsholtzia stauntonii 81
Enkianthus campanulatus 81
Epilobium angustifolium 46,
    4,37,57,138
Epilobium spicatum 46
Eranthis hyemalis 47
Erica 81,8,82
Erica carnea 82
Erica cinerea 82
Erica tetralix 81
Erigeron speciosus 47
Eryngium giganteum 47
Eryngium maritimum 47
Erysimum allionii 20
Erysimum cheiri 20
Erysimum hieracifolium 20
Eschscholzia californica 23
Euonymus alata 82
Eupatorium 4,47
Eupatorium perfoliatum 47
Euphorbia marginata 122,124
Evodia daniellii 82,4
Fagopyrum 4
Fagopyrum esculentum 23,24,138
Fagopyrum tataricum 24
Fagopyrum vulgare 23
Fagus sylvatica 82,121
Foeniculum officinale 48
Foeniculum vulgare 48
Fuchsia magellanica 82
Gaillardia 7
Gaillardia pulchella 24
Galanthus nivalis 48
Gaultheria shallon 83
Gaylussacia 4
Gaylussacia baccata 83
Gaylussacia brachycera 83
Gelsemium sempervirens 83,125
Geranium ibericum 48
Geranium phaeum 48
Geranium pratense 48
Geranium sanguineum 48

Geum 48
Gilia capitata 24
Gladiolus 48
Gleditsia triacanthos 83,2
Grossularia uva-crispa 100
Gypsophila paniculata 49
Gypsophila repens 49
Hamamelis mollis 83
Hamamelis virginiana 83
Hebe 84
Hedera helix 49
Hedysarum coronarium 49
Helenium autumnale 49,118
Helenium hoopesii 125
Helenium tenuifolium 118
Helianthus 24,5,50,138
Helianthus annuus 25,120
Helianthus orgyalis 50
Helianthus petiolaris 25,7
Helianthus salicifolius 50
Heliotropium curassavicum 25
Helleborus corsica 50
Helleborus lividus 50
Helleborus niger 50
Helleborus orientalis 50
Hemizonia 5
Heuchera sanguinea 50
Hyacinthus orientalis 50
Hydrangea 84
Hydrophyllum virginianum 50
Hyoscyamus niger 125
Hypericum 84
Hypericum perforatum 125
Hyssopus anisatus 36
Hyssopus officinalis 50
Hyssopus vulgaris 50
Iberis umbellata 25
Ilex aquifolium 84
Ilex glabra 84
Ilex opaca 84
Ilex verticillata 85
Impatiens balsamina 25
Impatiens glandulifera 25
Isatis tinctoria 26
Jacobaea elegans 32
Juglans mandshurica 125
Kalmia angustifolia 85,5
Kalmia latifolia 85,87,99,122,125
Koelreuteria paniculata 85
Labiatae 7,91
Larix decidua 85,121
Larix larcina 86
Lathyrus sylvestris 51
Lathyrus tuberosus 51
Laurus nobilis 86

Lavandula angustifolia 86
Lavandula × aurigerana 86
Lavandula × burnati 86
Lavandula × feraudi 86
Lavandula × guilloni 86
Lavandula × hortensis 86
Lavandula × hybrida 86
Lavandula × intermedia 86
Lavandula × leptostachya 86
Lavandula officinalis 86
Lavandula pyrenaica 86
Lavandula × senneni 86
Lavandula spica 86
Lavandula vera 86
Lavatera 26
Lavatera alba 26
Lavatera rosea 26
Lavatera thuringiaca 51
Lavatera trimestris 26
Ledum glandulosum 86
Ledum groenlandicum 86,5
Ledum palustre 86,122,125
Leguminosae 7
Leonurus cardiaca 51
Leonurus glaucescens 51
Leonurus quinquelobatus 51
Leonurus sibericus 26
Lespedeza bicolor 87
Lespedeza cryobotria 87
Lespedeza thunbergii 87
Leucojum vernum 51
Leucothoe 87,122,125
Leucothoe fontanesiana 87
Liatris pychnostchya 52
Liatris spicata 52
Ligustrum 5,117
Ligustrum japonicum 87
Ligustrum vulgare 87
Limnanthes douglasii 26
Limonium carolinianum 52
Limonium latifolium 52,26
Limonium sinuatum 26
Limonium vulgare 52
Linum flavum 52
Linum perenne 52
Linum usitatissimum 27,52
Liriodendron tulipifera 88,120
Lobularia maritima 27
Lonicera 88,6,7,80,89
Lonicera involucrata 88
Lonicera morrowii 88
Lonicera × purpusii 88
Lonicera standishii 88
Lonicera tatarica 88,7
Lonicera xylosteum 89

SCIENTIFIC NAMES

Lophanthus anisatus 36
Lotus 5
Lotus corniculatus 52,4,138
Lupinus 53
Lupinus angustifolia 53
Lupinus nootkatensis 53
Lupinus perennis 53
Lupinus polyphyllus 53
Lythrum salicaria 53
Magnolia 89
Magnolia acuminata 89
Magnolia grandiflora 89,118
Magnolia heptapeta 89
Magnolia quinquepeta 89
Magnolia stellata 89
Magnolia × soulangiana 89
Mahonia aquifolium 89,74
Majorana hortensis 30
Majorana majorana 30
Majorana vulgaris 30
Malus 90,4,63,91,93,94,97,155
Malus baccata 90
Malus communis 90,91
Malus coronaria 90
Malus domestica 90,91,120
Malus pumila 91,7
Malus sylvestris 90,91
Malva moschata 53
Malva thuringiaca 51
Marrubium vulgare 53
Medicago 28
Medicago falcata 54
Medicago intermedia 54
Medicago lupulinas 27
Medicago sativa 54,4,7,57,120
Medicago silvestris 54
Medicago subfalcata 54
Melilotus 27,4,28,29,36,51,52,57,138
Melilotus alba 28,7,27,29,56
Melilotus alba 'Annua' 28
Melilotus indica 27
Melilotus officinalis 29,7,25,27,28
Melissa officinalis 54
Mentha 55,56
Mentha aquatica 55
Mentha arvensis 55
Mentha longifolia 56
Mentha × piperita 55
Mentha pulegium 55
Mentha requienii 55
Mentha spicata 56,55
Monarda didyma 56
Monarda punctata 56
Muscari botryoides 56
Myosotis scorpioides 56

Myrica gale 91
Napus oleifera 17
Nemophila menziesii 29
Neocleome spinosa 20
Nepeta cataria 56,7,19
Nepeta × faassenii 56
Nepeta mussinii 56
Nicotiana affinia 29
Nicotiana alata 29
Nicotiana tabacum 29,125
Ocimum basilicum 29
Oenothera 57
Onobrychis 5
Onobrychis viciifolia 57,17
Opuntia 57
Opuntia polyacantha 57
Origanum majorana 30
Origanum vulgare 57,30
Oxydendrum aboreum 91
Paeonia 58
Papaver 8
Papaver orientale 58
Papaver rhoeas 30
Papaver somniferum 58,125
Parthenocissus quinquefolia 58
Paulownia tomentosa 91
Penstemon barbatus 58
Penstemon grandiflorus 58
Perovskia atriplicifolia 91
Persica 97
Persica vulgaris 95
Phacelia 51
Phacelia campanularia 30
Phacelia tanacetifolia 30
Phacelia viscida 30
Phaseolus 30,31,34,155
Phaseolus aureus 34
Phaseolus coccineus 31
Phaseolus lunatus 30,155
Phaseolus radiatus 34
Phellodendron amurense 92,125
Phyllodoce 75
Physostegia virginiana 58
Picea 8
Picea abies 92,121
Picea excelsa 92
Pieris 122,125
Pieris floribunda 92
Pieris japonica 92
Pimpinella anisum 31
Pinus 8,92
Pinus sylvestris 92,121
Polemonium caerulum 59
Polygonum 5,59
Polygonum amplexicaule 59

Polygonum aubertii 59
Polygonum cuspidatum 59
Polygonum fagopyrum 23
Polygonum sachalinese 59
Polygonum saggittatum 23
Polygonum tataricum 23,24
Populus 93,5,8,121
Populus tremuloides 93
Prunus 93,4,6,8,50,57,76,90,94,95,96,97
Prunus americana 93
Prunus armeniaca 94,93,155
Prunus avium 94,93
Prunus besseyi 93
Prunus cerasifera 93
Prunus cerasus 93
Prunus depressa 93
Prunus domestica 94,93
Prunus dulcis 94,155
Prunus incisa 93
Prunus laurocerasus 95
Prunus mahaleb 93
Prunus nigra 94
Prunus padus 95
Prunus pensylvanica 93
Prunus persica 95,93
Prunus pumila 94
Prunus sargentii 95,93
Prunus serotina 94
Prunus serrulata 95,93
Prunus spinosa 95
Prunus subhirtella 93
Prunus susquehanae 94
Prunus virginiana 96,93
Prunus yedoensis 96,93
Pseudotsuga glauca 96
Pseudotsuga menziesii 96,85,93,119,121
Ptelea trifoliata 97
Pulsatilla patens 38
Pycnanthemum virginianum 59
Pyracantha coccinea 97
Pyrus 97,57,90,93,94
Pyrus aucuparia 106
Pyrus baccata 5
Pyrus communis 97
Pyrus malus 90
Pyrus pumila 91
Quercus 98,5,125
Quercus alba 98
Quercus marcrocarpa 98
Quercus robur 98,121
Quercus rubra 98
Ranunculus 59,125
Raphanus campestris 18
Raphanus raphanistrum 5
Reseda odorata 31

Reynoutria japonica 59
Rhamnus 4
Rhamnus cathartica 98
Rhamnus frangula 98
Rhamnus purshiana 98,4
Rhododendron 99,87
Rhododendron arboreum 99,125
Rhododendron lutea 99,125
Rhododendron occidentalis 125
Rhododendron ponticum 99,122,125
Rhododendron prattii 99,125
Rhododendron thomsonii 99,126
Rhus glabra 99
Rhus typhina 99,5
Ribes 94
Ribes aureum 99,100
Ribes grossularia 100
Ribes odoratum 100
Ribes rubrum 100
Ribes sanguineum 100
Ribes uva-crispa 100
Robinia hispida 100
Robinia pseudoacacia 100,5,120
Robinia pseudoacacia 'Semperflorens' 101
Rosa 101,7,102
Rosa blanda 101
Rosa eglanteria 101
Rosa hugonis 102
Rosa johannensis 101
Rosa multiflora 102
Rosa nitida 101
Rosa rousseauiorum 101
Rosa rugosa 102
Rosa setigera 102
Rosa tomentosa 101
Rosa wichuraiana 102
Rosaceae 90
Rosmarinus officinalis 102
Rubus 103,4,5,120,156
Rubus deliciosus 103
Rubus fruticosus 103,156
Rubus idaeus 103
Rubus laciniatus 103
Rubus occidentalis 103
Rubus odoratus 104
Rubus parviflorus 104
Rudbeckia 31,4,60
Rudbeckia hirta 31
Rudbeckia laciniata 60
Salix 104,5,8,105
Salix alba 104,120
Salix bebbiana 104
Salix caprea 104,120
Salix cinerea 104
Salix discolor 105

Salix fragilis 104
Salix nigra 105
Salix purpurea 105
Salix sericea 104
Salix × smithiana 104
Salix triandra 104
Salsola 5
Salvia 60,32,61
Salvia hortensis 61
Salvia mellifera 60
Salvia nemorosa 60
Salvia officinalis 60
Salvia pratensis 60
Salvia sclarea 31
Salvia splendens 32
Salvia × superba 61
Salvia virgata nemorosa 60
Sambucus canadensis 4
Sapindus marginatus 126
Satureja 32,61
Satureja hortensis 32,61
Satureja montana 61,32
Scabiosa 32
Scabiosa atropurpurea 32,126
Scabiosa caucasica 61
Scilla siberica 61
Scrophularia 61
Scrophularia marilandica 61,2
Sedum spectabile 61
Senecio elegans 32
Senecio greyii 105
Senecio jacobaea 32,117
Sherperdia argentea 105
Sidalcea candida 61
Sinapis 8
Sinapis alba 32,18
Sinapis nigra 17
Solanum nigrum 122,126
Solidago 62,4,39,63,138
Solidago altissima 62
Solidago canadensis 62,7
Solidago lepida 62
Solidago occidentalis 63,7,62
Sophora davidii 106
Sophora japonica 106,126
Sophora viciifolia 106
Sorbus americana 106
Sorbus aria 106
Sorbus aucuparia 106
Sorbus intermedia 106
Stachys arvensis 126
Stachys byzantina 63
Stachys foeniculum 36
Stachys lanata 63
Stachys olympica 63

Statice armerica 38
Statice maritima 38
Statice sinuatum 26
Symphoricarpos albus 106,107
Symphoricarpos occidentalis 107,5,7
Symphoricarpos racemosus 106
Symphoricarpos rivularis 107
Syringa 107
Syringa amurense 126
Tagetes 33
Tamarix 107,118
Tamarix gallica 107
Tamarix parviflora 118
Tamarix pentandra 107
Tamarix ramosissima 107,118
Taraxacum 63,4,8
Taraxacum ambigens 64
Taraxacum ceratophorum 63
Taraxacum dumetorum 64
Taraxacum erythrosperum 64
Taraxacum lacerum 64
Taraxacum lapponicum 64
Taraxacum latilobum 64
Taraxacum laurentianum 64
Taraxacum officinale 63,7,57
Taraxacum phymatocarpum 63
Taxus 126
Taxus baccata 108
Taxus canadensis 108
Thalictrum 64
Thymus 32,61
Thymus serpyllum 64
Thymus vulgaris 64
Tilia 108,16,17,22,43,64,71,
    78,109,110,111,122,126,138
Tilia americana 109,4,108
Tilia amurensis 109
Tilia cordata 110,109,120
Tilia × euchlora 110,109
Tilia × europaea 110,109,111,120
Tilia japonica 109
Tilia maximowicziana 109
Tilia mongolica 110,109
Tilia orbicularis 109
Tilia parviflora 110
Tilia petiolaris 108.109
Tilia platyphyllos 110,109,120
Tilia tomentosa 111,109,120
Trifolium 64,8,22,33,34,
    65,66,79,117
Trifolium hybridum 65,4,7,57,138
Trifolium incarnatum 33,7,30,64
Trifolium pratense 65,4,7,28,30,33
    64,120,138

Trifolium repens **66**,4,7,16,18,33,57,65,106,138
Trifolium resupinatum 4
Trifolium serphyllum 64
Trillium **66**
Tulipa 126
Tulipa kaufmaniana **66**
Ulex europaeus 111
Ulmus 4,8
Ulmus americana 111
Vaccinium 111,4,77,112
Vaccinium angustifolium 112
Vaccinium arboreum 112
Vaccinium corymbosum 112,156
Vaccinium macrocarpon 112
Vaccinium myrtillus 111
Vaccinium pensylvanicum 112
Vaccinium uliginosum 111
Vaccinium vitis-idaea 113
Valeriana officinalis **66**
Valeriana olitoria **66**
Vassinium uliginosum 111
Veratrum 126
Verbascum blatteria 33
Verbascum thapsus 33
Verbesina alternifolia **67**,2
Veronica spicata **67**
Viburnum 113
Viburnum opulus 113
Viburnum plicatum 113
Viburnum prunifolium 113
Vicia 5,6,34
Vicia faba 33,120,155
Vicia pannonica 34
Vicia sativa 34
Vicia villosa 34
Vigna aureus 30
Vigna radiate 34
Vitex incisa 113
Vitex negundo 113
Vitis parthenocissus quinquefolia 58
Vitis vulpina **67**
Weigela florida 113
Wisteria chinensis 114
Wisteria floribunda 113
Wisteria sinensis 114
Yucca filamentosa 114
Yucca glauca 114
Zea mays 34,4,121
Zigadenus venenosus 126
Zinnia elegans 35

17. MAPS

Overleaf

FIG. 2. Map of plant hardiness zones of Canada (extracted from Cole, T.J. 1979. A Checklist of Ornamental Trees for Canada, *Publ. 1343, Agric. Can., Ottawa*. Reproduced with permission of Supply and Services Canada.)

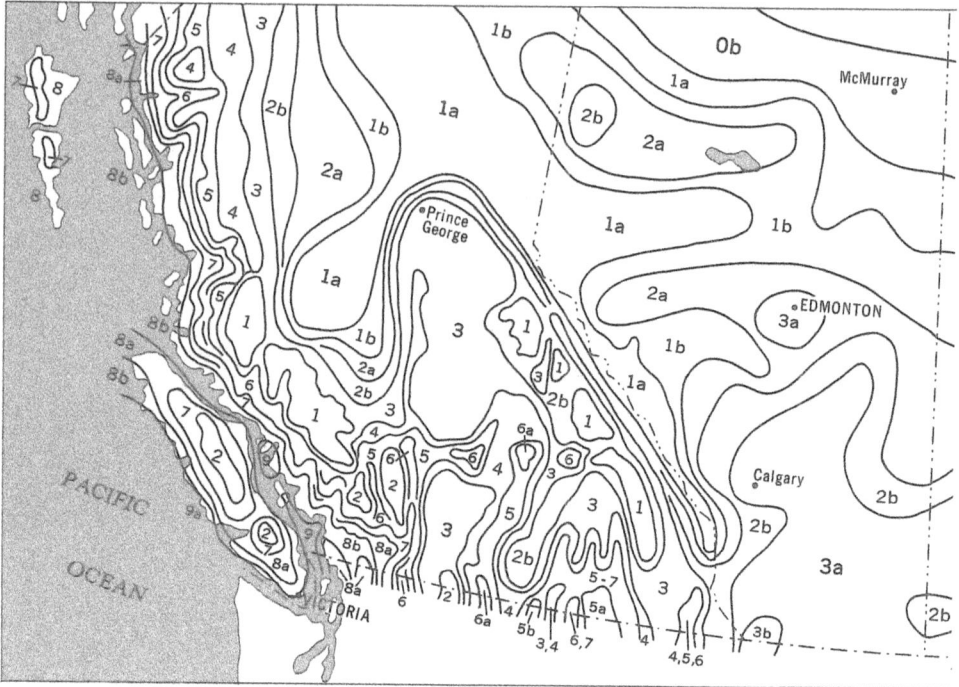

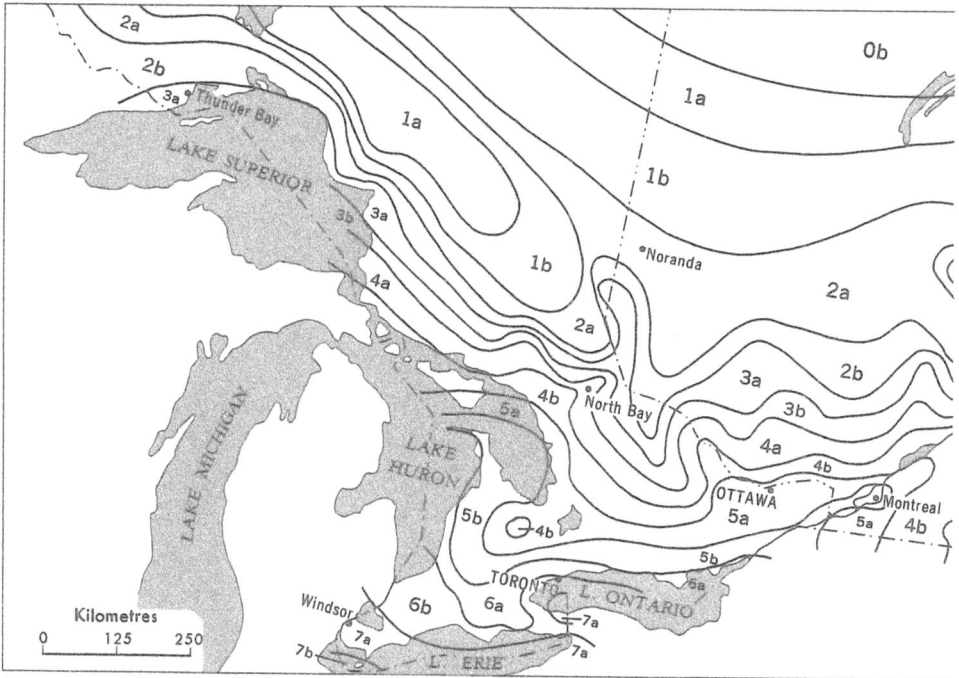

Plant hardiness zones in Canada.

MAPS

Oa

Oa

Ob

Ob

Flin
Flon

1a

1a

1b

Prince Albert

1b

Lake Winnipeg

1b

2a

1a

3a

2a

Saskatoon

1b

2a

2b

1b

1a

2a

WINNIPEG
3a

2b

3a
Kenora

•REGINA

2b

Brandon

3a

2b
Thunder Bay
3a

3a

2b

3b

WESTERN CANADA

1a

1b

2a

2b

3b

St. Lawrence R.

3a

3b
4a
4a

3b

3b

3b

3a
4a

4b

5b

5a

3a
3b

4b

4a

3a

4b

5b

6a

5b

QUEBEC

2

3a

5a

5a
6a

4a

4b

6a

5b

4a

5b

ATLANTIC OCEAN

4b

3b

5a
5b

5a

HALIFAX

Saint
John

5b

5b

6a

6b

EASTERN CANADA

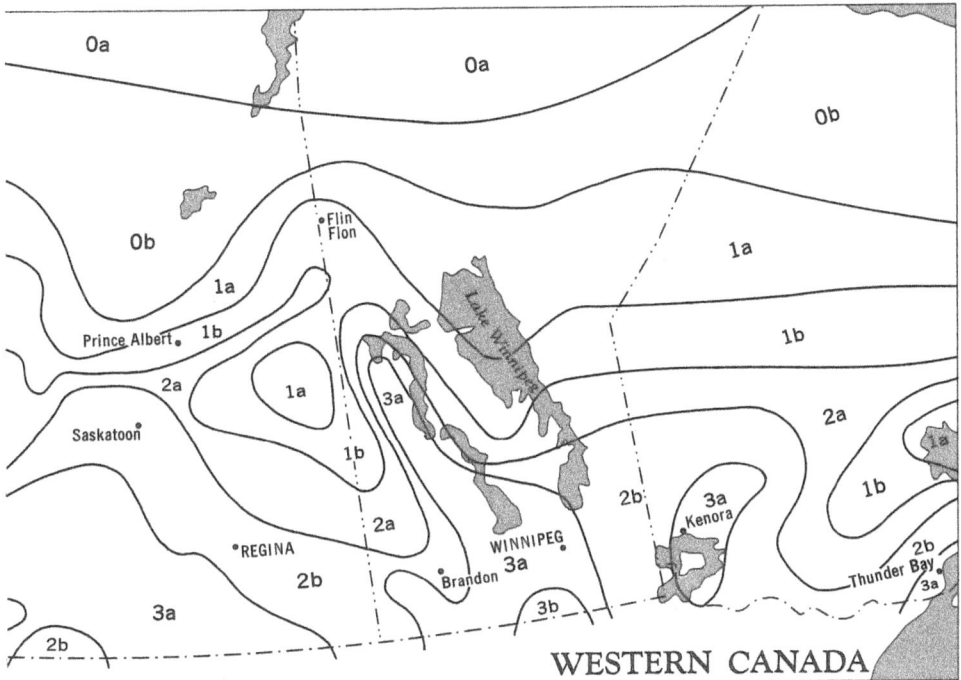

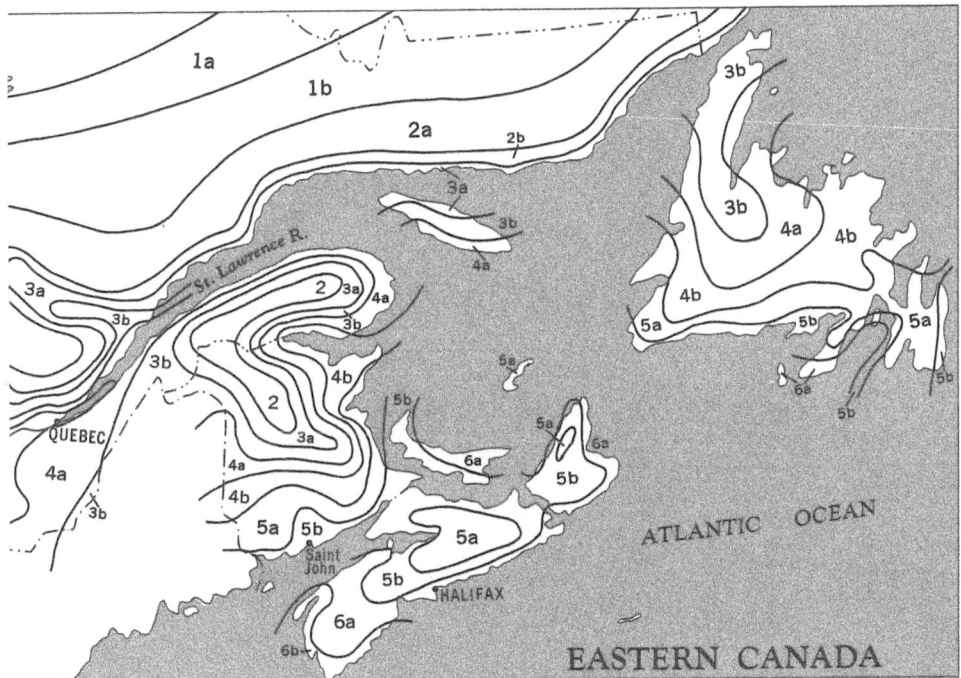

Cartography by the Land Resource Research Institute, Research Branch, Agriculture Canada, 1980.

MAPS

197

Hudson Bay
MAN
SASK.
ALB
BC
ONT
QUE
LAB
NFLD
N.B.
P.E.I.
N.S.
ME
V.T.
N.H.
MASS
R.I.
CONN
N.Y.
N.J.
PA
MD
DEL
VA
W.VA
N. CAR
S. CAR
GA
FLA
ALA
MISS
LA
TENN
KY
OHIO
IND
ILL
MICH
WISC
MINN
IOWA
MO
ARK
TEX
OKLA
KAN
NEB
S. DAK
N. DAK
MON
WYO
COL
NEW MEX
ARIZ
UTAH
NEV
IDAHO
ORE
WASH
CAL
MEXICO
KEY
AVERAGE ANNUAL MINIMUM TEMPERATURES FOR EACH ZONE
Zone °C °F
1 below -46 below -50
2 - 37 to - 46 -35 to -50
3 - 29 to -37 -20 to -35
4 - 23 to -29 -10 to -20
5 - 21 to -23 - 5 to -10
6 - 15 to - 21 5 to - 5
7 - 12 to - 15 10 to 5
8 - 7 to -12 20 to 10
9 - 1 to - 7 30 to 20
10 4 to - 1 40 to 30
Originally compiled by The Arnold Arboretum, Harvard University, Jamaica Plain, Mass., Jan. 30, 1971

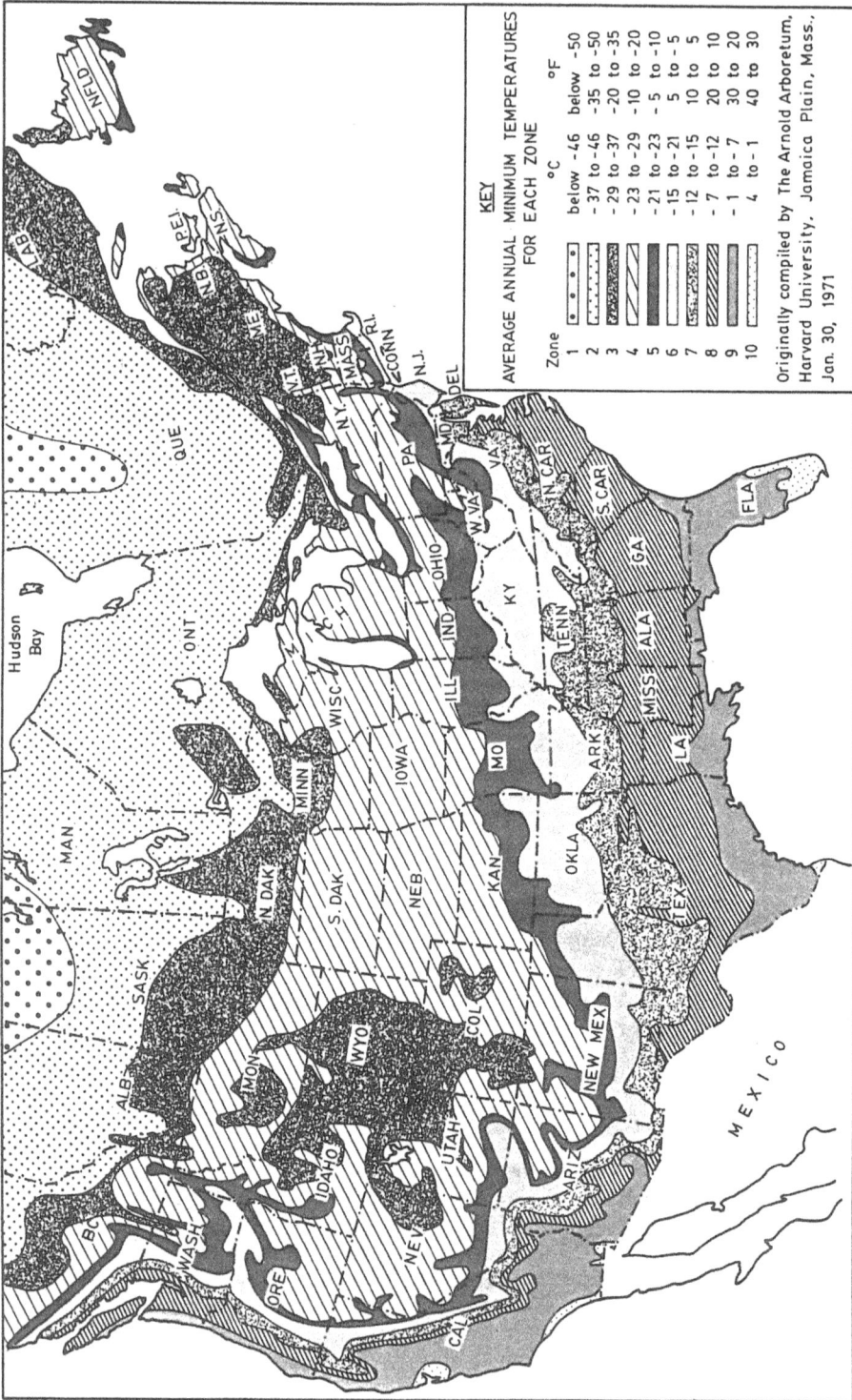

FIG. 3. Hardiness zones of the United States of America and Canada (extracted from Wyman 1971[95]. Redrawn by J. Ramsay 1987 and reproduced with permission from the Arboretum of Harvard University).

Field Notes

Field Notes

Field Notes

Field Notes

www.ingramcontent.com/pod-product-compliance
Lightning Source LLC
Chambersburg PA
CBHW041118280326
41928CB00060B/3455